NEW EDITION

WORKBOOK for New Progress in Mathematics

An innovative approach including *two* options:

- Pre-Algebra
- Algebra

Rose Anita McDonnell
Catherine D. LeTourneau
Anne Veronica Burrows

with

Dr. Elinor R. Ford

Sadlier-Oxford
A Division of William H. Sadlier, Inc.

Home Office: 9 Pine Street,
New York, NY 10005
ISBN 0-8215-1728-7
22 23 24 25 / 10 09 08 07

Table of Contents

Whole Numbers to Trillions*

Name ______________________

Date ______________________

Use the place value chart below to write the following numbers.

1. eleven million, six hundred forty thousand, eighteen
2. eighty-two billion, nine million, two thousand ten
3. ten trillion, five hundred million, two thousand

	Trillions period			Billions period			Millions period			Thousands period			Ones period		
	hundreds	tens	ones	hundreds	tens	ones	hundreds	tens	ones	hundreds	tens	ones	hundreds	tens	ones
1.															
2.															
3.															

In what place is the underlined digit? What is its value?

4. 2$\underline{4}$,679 ________

5. 1$\underline{0}$9,672,521 ________

6. $\underline{9}$4,276,541,327 ________

7. 6,846,9$\underline{7}$2,420 ________

8. 1$\underline{2}$5,473,890,367 ________

9. $\underline{2}$,075,398,123 ________

10. 63,520,$\underline{6}$41 ________

11. 9,$\underline{9}$99,999,999,001 ________

12. 5,382,4$\underline{9}$2,770 ________

Use these numbers to answer the following questions: **a.** 8,752,461,304,001
b. 662,491,307,262,859

13. The value of 5 in **a** is how many times greater than the value of 5 in **b**? ______________

14. The value of 3 in **a** is how many times less than the value of 3 in **b**? ______________

15. The value of 8 in **a** is how many times greater than the value of 8 in **b**? ______________

Solve.

16. About how old would you be if you lived until you were a billion hours old? ______________

17. About how many years have passed since a trillion hours ago? Was America discovered yet? Were the pyramids built? ______________

18. The current population of the United States is about 251,400,000 people. The annual budget is about $3 trillion. How much would this be per person in the country? ______________

*Use with Lesson 1-1, text pages 2–3.

Reading and Writing Whole Numbers and Decimals*

Name ______________________

Date ______________________

Read each number. Give the value of the circled digit.

1. 34.2⓪676 ____________

2. 98.5050① ____________

3. 118.⑦4213 ____________

4. 2.②2222 ____________

5. 0.69138⑦ ____________

6. 442.922④26 ____________

Write in standard form.

7. $(8 \times 10{,}000) + (4 \times 100) + (2 \times 1)$ ____________

8. $(5 \times 100{,}000) + (3 \times 1000) + (6 \times 1)$ ____________

9. $(7 \times 0.1) + (8 \times 0.0001)$ ____________

10. $(5 \times 0.1) + (9 \times 0.0001)$ ____________

Write the standard numeral.

11. 6 million ____________

12. 82 billion ____________

13. 7 billion ____________

14. 95 trillion ____________

15. 4.2 billion ____________

16. 5.1 million ____________

17. 3.8 trillion ____________

18. 7.75 billion ____________

Write the number that is named by the *scrambled* digits in exercises 19–21.

19. 8 millionths
0 thousandths
6 tenths
3 ones
5 ten thousandths

20. 2 thousands
9 millionths
9 ones
5 tenths
1 ten thousandth

21. 2 hundredths
6 billionths
9 ten thousandths
8 millionths
2 thousandths

22. Write the *least* number using the digits 0,9,8,7,6 each once. ____________

23. Write the *greatest* number using the digits 0,9,8,7,6 each once. ____________

24. Write the number that is exactly 5,000,000 *less* than 9,125,799. ____________

25. Write the number that is 0.012345 *greater* than 1.070707. ____________

*Use with Lesson 1-1, text pages 2–3.

Adding and Subtracting Whole Numbers and Decimals*

Name ______________________

Date ______________________

Find the sum or difference.

1.
0.3714
0.2401
0.096
0.4412
+ 0.55

2.
0.764
0.4036
0.5271
0.321
+ 0.908

3.
0.2
0.364
0.101
0.8964
+ 0.9

4.
0.48
0.6801
0.725
0.655
+ 0.861

5.
10.27
− 4.38

6.
27.82
− 9.47

7.
47.801
− 9.38

8.
702.08
− 96.4

9. 0.41 + 0.328 + 0.92 + 0.1435 = ______________

10. 38 − 4.921 = ______________

11. 0.8654 + 0.9284 + 0.91 = ______________

12. 5.7 − 0.346 = ______________

13. 5.1023 + 1.06 + 2.007 + 6.8 = ______________

14. 18.1 − 9.341 = ______________

Estimate by rounding each number to the nearest tenth.

15.
0.324
0.9281
0.24
+ 0.8361

16.
1.207
− 0.46

17.
0.24
0.3416
0.291
0.834
+ 0.33

Estimate by rounding each number to the nearest whole number.

18.
27.046
− 13.9

19.
824.06
− 421.73

20.
38.53
− 18.49

***Use with Lessons 1-4, 1-5, text pages 8–11.**

Multiplying Whole Numbers*

Name ____________________
Date ____________________

Multiply.

1. $\begin{array}{r} 8261 \\ \times\ 438 \\ \hline \end{array}$

2. $\begin{array}{r} 9063 \\ \times\ 71 \\ \hline \end{array}$

3. $\begin{array}{r} 6428 \\ \times\ 207 \\ \hline \end{array}$

4. $\begin{array}{r} 3498 \\ \times\ 276 \\ \hline \end{array}$

5. $\begin{array}{r} 4307 \\ \times\ 691 \\ \hline \end{array}$

6. $\begin{array}{r} 8041 \\ \times\ 257 \\ \hline \end{array}$

7. $\begin{array}{r} 5476 \\ \times\ 355 \\ \hline \end{array}$

8. $\begin{array}{r} 6010 \\ \times\ 305 \\ \hline \end{array}$

9. $\begin{array}{r} 2041 \\ \times\ 920 \\ \hline \end{array}$

10. $\begin{array}{r} 5376 \\ \times\ 208 \\ \hline \end{array}$

11. $\begin{array}{r} 9008 \\ \times\ 467 \\ \hline \end{array}$

12. $\begin{array}{r} 3276 \\ \times\ 891 \\ \hline \end{array}$

Estimate by rounding each number to its greatest place-value position.

13. 6041 × 935 ____________

14. 888 × 222 ____________

15. 782 × 43 ____________

16. 8965 × 423 ____________

17. 12,423 × 9 ____________

18. 53,219 × 452 ____________

19. 302,009 × 52 ____________

20. 87,526 × 347 ____________

21. 472,351 × 63 ____________

22. 92,718 × 586 ____________

*Use with Lesson 1-6, text pages 12–13.

Multiplying Decimals*

Name ______________________

Date ______________________

Find the product.

1. 23.4×8.6

2. 4.076×3.9

3. 421.7×0.076

4. 8.71×0.421

5. 80.63×9.07

6. 24.6×3.82

7. 5.071×0.34

8. 7.031×2.7

9. 4.075×3.12

10. 8.076×0.5

11. 4.071×3.7

12. 72.51×0.08

13. 8.055×9.3

14. 84.75×1.05

15. 3.067×9.1

16. 6.934×0.8

Estimate by rounding each number to its greatest place-value position.

17. 43.6×8.9 ______________________

18. 59.09×21.6 ______________________

19. 0.086×3.5 ______________________

20. 32.4×0.91 ______________________

21. 80.6×1.7 ______________________

22. 302.4×0.079 ______________________

Solve.

23. Find the cost of 11.3 meters of plastic tubing if one meter costs $.72. ______________________

24. Find the weight of 4.65 meters of copper tubing if one meter weighs 1.2 kg. ______________________

***Use with Lesson 1-6, text pages 12–13.**

Dividing Whole Numbers*

Name ______________________

Date ______________________

Divide.

1. $63\overline{)49,140}$ **2.** $54\overline{)24,300}$ **3.** $91\overline{)820,850}$ **4.** $86\overline{)68,807}$

5. $73\overline{)29,721}$ **6.** $58\overline{)43,500}$ **7.** $92\overline{)55,209}$ **8.** $65\overline{)63,245}$

9. $46\overline{)43,746}$ **10.** $75\overline{)73,650}$ **11.** $85\overline{)57,800}$ **12.** $76\overline{)63,769}$

13. $103\overline{)3720}$ **14.** $259\overline{)21,497}$ **15.** $481\overline{)32,708}$ **16.** $607\overline{)126,260}$

17. $570\overline{)263,977}$ **18.** $349\overline{)16,098}$ **19.** $462\overline{)23,123}$ **20.** $709\overline{)19,800}$

Estimate by rounding each number to its greatest place-value position.

21. $2967 \div 62$ ______________ **22.** $1904 \div 36$ ______________

23. $38,416 \div 47$ ______________ **24.** $356,124 \div 18$ ______________

25. $423 \div 15$ ______________ **26.** $76,032 \div 413$ ______________

*Use with Lesson 1-7, text pages 14–15.

Dividing Decimals*

Name ______________________

Date ______________________

Find the quotient.

1. $9\overline{)1.44}$
2. $2\overline{)3.46}$
3. $24\overline{)81.6}$
4. $29\overline{)72.413}$
5. $0.9\overline{)18.27}$
6. $0.02\overline{)16.43}$
7. $1.5\overline{)25.305}$
8. $0.07\overline{)217.42}$

Find the quotient to the nearest hundredth.

9. $2.3\overline{)46.07}$
10. $4.8\overline{)95.41}$
11. $3.4\overline{)6.317}$
12. $5.06\overline{)94.84}$

Estimate by rounding each number to its greatest place-value position.

13. $0.39 \div 8 =$ ______________
14. $8.51 \div 2.71 =$ ______________
15. $24.14 \div 0.41 =$ ______________
16. $4.38 \div 0.95 =$ ______________
17. $5.68 \div 2.94 =$ ______________
18. $3.91 \div 0.39 =$ ______________

Solve.

19. How much paint is required to paint one window frame if 18.45 L are needed for 45 frames? ______________

20. Which costs less per kilogram: 3.25 kg of buttons for $2.11 or 4.45 kg of metal snaps for $3.65? ______________

*Use with Lesson 1-8, text pages 16 – 17.

Estimating Whole Numbers and Decimals*

Name ______________________

Date ______________________

Compare. Use $<$, $=$, or $>$.

1. 400×90 ______ 40×900
2. 6×600 ______ 60×60
3. 300×700 ______ 2.1×1000
4. 0.5×52.2 ______ 1000×0.522
5. 0.487×500 ______ 0.5×487
6. $9.09 \div 30$ ______ $30 \div 9.09$

Estimate the sum or difference.

7. $47{,}300 + 25{,}620 + 24{,}100 + 10{,}520$ ______________________
8. $6.96 + 12.22 + 8.65 + 9.01 + 3.25$ ______________________
9. $13.001 + 5.021 + 4.225 + 8.002$ ______________________

Estimate the product or quotient.

10. 840×21 ______
11. $14{,}268 \times 39$ ______
12. 10.6×5.8 ______
13. $6448 \div 31$ ______
14. $51{,}416 \div 18$ ______
15. $36.8 \div 4.2$ ______

Find unreasonable answers. Write R for reasonable; U for unreasonable.
Estimate. Do not compute.

16. $1{,}411 + 3{,}656 + 7{,}722 \approx 13{,}000$ ______
17. $4.8 + 3.9 + 7.2 + 0.8 \approx 16$ ______
18. $72{,}328 \div 7.7 \approx 8200$ ______
19. $426 \times 34.5 \approx 15{,}000$ ______

Solve.

20. Create a division problem with whole numbers to fit these descriptions.

    a. A 5-digit number is divided by a 3-digit number. The remainder is 124. ______________

    b. A 4-digit number is divided by a 3-digit number. The remainder is 60. ______________

    c. A 6-digit number is divided by a 3-digit number. The remainder is 30. ______________

*Use with Lesson 1-9, text pages 18–19.

Number Properties*
(And Computational Shortcuts)

Name ______________________

Date ______________________

Match each example with the number property used to solve it. Use each example only once.

1. _______ Identity for multiplication	**a.** $341{,}296 + [49 - (7 \times 7)]$
2. _______ Commutative for addition	**b.** $86 + 95 + 814 + 505 = 86 + 814 + 95 + 505$
3. _______ Commutative and associative for multiplication	**c.** $(8 \times 49) + (38 \times 49) = (8 + 38) \times 49$
4. _______ Associative for addition	**d.** $92 \times 10 \times 48 \times 2 = (94 \times 10) \times (48 \times 2)$
5. _______ Commutative and identity for addition	**e.** $(9 \div 9) \times 3247$
6. _______ Distributive	**f.** $12 + 8 + 24 + 64 = (12 + 8) + (24 + 64)$
7. _______ Commutative for multiplication	**g.** $8.6 + 921 - 8.6 = 8.6 - 8.6 + 921$
8. _______ Identity for addition	**h.** $92 \times 43 \times 2 \times 20 = 92 \times 2 \times 43 \times 20 = (92 \times 2) \times (43 \times 20)$
9. _______ Commutative and associative for addition	**i.** $194 + 87 + 106 + 413 = 194 + 106 + 87 + 413 = (194 + 106) + (87 + 413)$
10. _______ Associative for multiplication	**j.** $8 \times 43 \times 21 = 8 \times 21 \times 43$

Solve. Use properties wherever possible.

11. $(7 \times 28) + (7 \times 42) =$ ______________

12. $2.5 \times 813 \times 2 =$ ______________

13. $4.2 + 3.7 + 0.8 =$ ______________

14. $4 \times 0.25 \times 847 =$ ______________

15. $46 + 92 + 31 + 54 + 69 + 8 =$ ______

16. $8.5 + [9.23 - (5.41 - 3.82)] =$ __________

Compare. Write $<$, $=$, or $>$.

17. 20×20 _____ 400

18. $(1.4 \times 0.4) + (1.4 \times 0.2)$ _____ (1.4×0.9)

19. 2.83×2.61 _____ 2.61×2.83

20. 3.41×100 _____ 0.341×10

21. 3×120 _____ 40×90

22. 470×50 _____ 47×500

*Use with Lesson 1-10, text pages 20–21.

Fractions and Mixed Numbers*

Name ______________________

Date ______________________

Change each to a fraction.

1. $5\frac{1}{3}$ = ______ **2.** $7\frac{2}{5}$ = ______ **3.** $8\frac{1}{9}$ = ______ **4.** $6\frac{1}{4}$ = ______

5. $3\frac{2}{3}$ = ______ **6.** $4\frac{2}{9}$ = ______ **7.** $2\frac{1}{8}$ = ______ **8.** $12\frac{1}{10}$ = ______

9. $14\frac{1}{5}$ = ______ **10.** $2\frac{3}{7}$ = ______ **11.** $9\frac{3}{5}$ = ______ **12.** $3\frac{1}{4}$ = ______

13. $13\frac{2}{5}$ = ______ **14.** $8\frac{3}{11}$ = ______ **15.** $16\frac{2}{3}$ = ______ **16.** $8\frac{3}{9}$ = ______

17. $20\frac{5}{6}$ = ______ **18.** $9\frac{10}{11}$ = ______ **19.** $12\frac{4}{5}$ = ______ **20.** $10\frac{5}{13}$ = ______

21. $16\frac{4}{5}$ = ______ **22.** $8\frac{2}{7}$ = ______ **23.** $12\frac{3}{8}$ = ______ **24.** $20\frac{2}{3}$ = ______

Change each to a mixed number. (Show the fraction part in simplest form.)

25. $\frac{29}{3}$ = ______ **26.** $\frac{89}{5}$ = ______ **27.** $\frac{17}{3}$ = ______ **28.** $\frac{9}{7}$ = ______

29. $\frac{36}{7}$ = ______ **30.** $\frac{41}{9}$ = ______ **31.** $\frac{82}{7}$ = ______ **32.** $\frac{48}{5}$ = ______

33. $\frac{31}{6}$ = ______ **34.** $\frac{53}{10}$ = ______ **35.** $\frac{49}{5}$ = ______ **36.** $\frac{75}{8}$ = ______

37. $\frac{52}{7}$ = ______ **38.** $\frac{39}{4}$ = ______ **39.** $\frac{28}{3}$ = ______ **40.** $\frac{32}{5}$ = ______

41. $\frac{91}{4}$ = ______ **42.** $\frac{88}{6}$ = ______ **43.** $\frac{148}{12}$ = ______ **44.** $\frac{175}{13}$ = ______

45. $\frac{162}{4}$ = ______ **46.** $\frac{189}{10}$ = ______ **47.** $\frac{215}{15}$ = ______ **48.** $\frac{420}{22}$ = ______

Solve.

49. José worked for $3\frac{3}{4}$ hours. Change this mixed number to a fraction. ______

50. Jessica pitched $\frac{10}{3}$ innings of the softball game. Change this fraction to a mixed number. ______

51. Heather used $2\frac{2}{3}$ cups of flour in her recipe. Change this mixed number to a fraction. ______

52. Zach is $\frac{65}{12}$ feet tall. Change this fraction to a mixed number. ______

*Use with Lesson 2-1, text pages 32–33.

Equivalent Fractions*

Name ______________________

Date ______________________

Complete to show equivalent fractions for each.

1. $\frac{1}{5} = \frac{\square}{10} = \frac{\square}{20} = \frac{\square}{30} = \frac{8}{\square}$

2. $\frac{2}{7} = \frac{\square}{14} = \frac{\square}{21} = \frac{10}{\square} = \frac{14}{\square}$

3. $\frac{2}{9} = \frac{\square}{18} = \frac{\square}{45} = \frac{16}{\square} = \frac{\square}{81}$

4. $\frac{7}{10} = \frac{\square}{20} = \frac{35}{\square} = \frac{\square}{100} = \frac{140}{\square}$

5. $\frac{3}{13} = \frac{\square}{26} = \frac{15}{\square} = \frac{\square}{104} = \frac{30}{\square}$

6. $\frac{3}{4} = \frac{\square}{20} = \frac{21}{\square} = \frac{\square}{36} = \frac{36}{\square}$

Complete each equivalent fraction.

7. $\frac{8}{9} = \frac{\square}{72}$

8. $\frac{4}{7} = \frac{\square}{21}$

9. $\frac{3}{5} = \frac{\square}{40}$

10. $\frac{2}{9} = \frac{\square}{63}$

11. $\frac{42}{49} = \frac{6}{\square}$

12. $\frac{3}{10} = \frac{6}{\square}$

13. $\frac{10}{15} = \frac{2}{\square}$

14. $\frac{12}{30} = \frac{2}{\square}$

15. $\frac{25}{50} = \frac{\square}{10}$

16. $\frac{4}{5} = \frac{28}{\square}$

17. $\frac{12}{48} = \frac{\square}{12}$

18. $\frac{55}{110} = \frac{1}{\square}$

Write each fraction in simplest form.

19. $\frac{40}{56}$ = ______

20. $\frac{16}{24}$ = ______

21. $\frac{8}{20}$ = ______

22. $\frac{18}{72}$ = ______

23. $\frac{15}{45}$ = ______

24. $\frac{7}{11}$ = ______

25. $\frac{81}{108}$ = ______

26. $\frac{14}{21}$ = ______

27. $\frac{14}{56}$ = ______

28. $\frac{20}{56}$ = ______

29. $\frac{49}{70}$ = ______

30. $\frac{19}{29}$ = ______

31. $\frac{28}{36}$ = ______

32. $\frac{54}{78}$ = ______

33. $\frac{48}{144}$ = ______

34. $\frac{18}{48}$ = ______

35. $\frac{36}{42}$ = ______

36. $\frac{16}{48}$ = ______

37. $\frac{27}{42}$ = ______

38. $\frac{24}{28}$ = ______

***Use with Lesson 2-2, text pages 34–35.**

Factors, Prime and Composite Numbers*

Name ____________________

Date ____________________

List the factors of each of the following numbers.

1. 14 ____________________ **2.** 21 ____________________

3. 35 ____________________ **4.** 25 ____________________

5. 8 ____________________ **6.** 75 ____________________

7. 15 ____________________ **8.** 39 ____________________

9. 30 ____________________

10. 36 ____________________

11. 60 ____________________

12. 72 ____________________

13. 120 ____________________

Find the prime numbers that complete each of these statements.

14. ______ + ______ = 5 **15.** ______ + ______ = 12 **16.** ______ + ______ = 61

Find the prime numbers between 100 and 200.

17. Use a calculator to find the numbers that are multiples of other numbers. Cross out these numbers on the chart. The numbers left are the prime numbers between 100 and 200. (Hint: You do not have to try a prime factor greater than 13.)

Prime Numbers

101	102	103	104	105	106	107	108	109	110
111	112	113	114	115	116	117	118	119	120
121	122	123	124	125	126	127	128	129	130
131	132	133	134	135	136	137	138	139	140
141	142	143	144	145	146	147	148	149	150
151	152	153	154	155	156	157	158	159	160
161	162	163	164	165	166	167	168	169	170
171	172	173	174	175	176	177	178	179	180
181	182	183	184	185	186	187	188	189	190
191	192	193	194	195	196	197	198	199	200

*Use with Lesson 2-3, text page 36.

Prime Factorization*

Name ______________________

Date ______________________

Draw a factor tree for each number. Then give the prime factorization for each.

1. 56 __________

2. 42 __________

Write the prime factorization of each number using exponents.

3. 44 __________

4. 32 __________

5. 18 __________

6. 72 __________

7. 120 __________

8. 105 __________

9. 20 __________

10. 65 __________

11. 90 __________

12. 180 __________

13. 48 __________

14. 78 __________

Write the composite number for each prime factorization.

15. $2^3 \times 3^3$ __________

16. $2 \times 3^3 \times 7$ __________

17. $3^2 \times 11$ __________

18. $2^2 \times 5^2$ __________

19. $2^3 \times 3^3 \times 7$ __________

20. $2^5 \times 5$ __________

21. $3 \times 7^2 \times 11$ __________

22. $2^4 \times 3^2$ __________

23. 3×5^2 __________

24. $3^2 \times 7^2$ __________

25. $3^2 \times 5 \times 13$ __________

26. $5 \times 11 \times 13$ __________

*Use with Lesson 2-4, text page 37.

Greatest Common Factor; Least Common Multiple*

Name ______________________

Date ______________________

Find the GCF.

1. 15 and 18 ______________
2. 56 and 16 ______________
3. 42 and 63 ______________
4. 70 and 28 ______________
5. 28 and 56 ______________
6. 35 and 15 ______________
7. 14 and 84 ______________
8. 30 and 45 ______________
9. 28 and 36 ______________
10. 14 and 49 ______________

Write the prime factorization in exponential form. Then write the GCF.

11. 9 ______________
    12 ______________
    GCF: ______________
12. 24 ______________
    18 ______________
    GCF: ______________
13. 8 ______________
    20 ______________
    GCF: ______________
14. 50 ______________
    75 ______________
    GCF: ______________

Find the LCM.

15. 12 and 8 ______________
16. 3 and 8 ______________
17. 6 and 9 ______________
18. 4 and 10 ______________
19. 5 and 6 ______________
20. 2 and 7 ______________
21. 8 and 10 ______________
22. 18 and 3 ______________
23. 9 and 5 ______________
24. 36 and 9 ______________

Write the prime factorization in exponential form. Then write the LCM.

25. 6 ______________
    8 ______________
    LCM: ______________
26. 20 ______________
    25 ______________
    LCM: ______________
27. 8 ______________
    12 ______________
    40 ______________
    LCM: ______________
28. 35 ______________
    60 ______________
    70 ______________
    LCM: ______________

*Use with Lessons 2-5, 2-6, text pages 38–39.

Renaming Fractions*

Name ______________________

Date ______________________

Use a power of ten to change each to a decimal.

1. $\frac{4}{50}$ = ________ **2.** $\frac{9}{25}$ = ________ **3.** $\frac{19}{20}$ = ________ **4.** $\frac{1}{10}$ = ________

5. $2\frac{1}{2}$ = ________ **6.** $4\frac{1}{25}$ = ________ **7.** $6\frac{1}{5}$ = ________ **8.** $7\frac{1}{20}$ = ________

Change each to a terminating or repeating decimal.

9. $\frac{2}{5}$ = ________ **10.** $\frac{7}{10}$ = ________ **11.** $\frac{5}{8}$ = ________ **12.** $\frac{3}{20}$ = ________

13. $\frac{5}{16}$ = ________ **14.** $\frac{2}{11}$ = ________ **15.** $\frac{4}{25}$ = ________ **16.** $\frac{3}{10}$ = ________

17. $\frac{7}{9}$ = ________ **18.** $\frac{3}{50}$ = ________ **19.** $\frac{3}{4}$ = ________ **20.** $\frac{2}{9}$ = ________

21. $2\frac{1}{8}$ = ________ **22.** $6\frac{1}{4}$ = ________ **23.** $1\frac{1}{3}$ = ________ **24.** $8\frac{1}{2}$ = ________

25. $9\frac{2}{15}$ = ________ **26.** $4\frac{3}{40}$ = ________ **27.** $7\frac{1}{6}$ = ________ **28.** $4\frac{2}{3}$ = ________

29. $2\frac{1}{9}$ = ________ **30.** $5\frac{3}{11}$ = ________ **31.** $6\frac{4}{9}$ = ________ **32.** $3\frac{4}{5}$ = ________

Change the fraction to a decimal.

33. Two and seven-eighths truckloads were filled with coal. ______________________

34. The board measured $\frac{15}{4}$ inches. ______________________

35. The fabric measured $10\frac{3}{10}$ yards. ______________________

*Use with Lesson 2-7, text pages 40–41.

Comparing and Ordering Fractions*

Name ______________________

Date ______________________

Compare. Write <, =, or >.

1. $\frac{1}{5}$ ______ $\frac{2}{9}$ **2.** $\frac{1}{3}$ ______ $\frac{2}{5}$ **3.** $\frac{1}{6}$ ______ $\frac{1}{5}$ **4.** $\frac{1}{8}$ ______ $\frac{2}{11}$

5. $\frac{3}{7}$ ______ $\frac{4}{9}$ **6.** $\frac{5}{8}$ ______ $\frac{7}{10}$ **7.** $\frac{8}{15}$ ______ $\frac{9}{20}$ **8.** $\frac{5}{6}$ ______ $\frac{7}{8}$

9. $\frac{4}{5}$ ______ $\frac{9}{10}$ **10.** $\frac{2}{11}$ ______ $\frac{7}{9}$ **11.** $\frac{3}{4}$ ______ $\frac{15}{19}$ **12.** $\frac{2}{3}$ ______ $\frac{11}{12}$

13. $1\frac{2}{3}$ ______ $1\frac{3}{5}$ **14.** $2\frac{1}{9}$ ______ $2\frac{3}{11}$ **15.** $8\frac{1}{3}$ ______ $8\frac{4}{15}$ **16.** $5\frac{1}{5}$ ______ $5\frac{4}{18}$

17. $\frac{12}{20}$ ______ $\frac{8}{15}$ **18.** $3\frac{1}{3}$ ______ $2\frac{4}{3}$ **19.** $\frac{5}{18}$ ______ $\frac{4}{15}$ **20.** $6\frac{1}{9}$ ______ $6\frac{4}{25}$

21. $9\frac{1}{8}$ ______ $9\frac{4}{7}$ **22.** $15\frac{2}{3}$ ______ $15\frac{5}{9}$ **23.** $12\frac{5}{6}$ ______ $10\frac{17}{6}$ **24.** $10\frac{3}{8}$ ______ $10\frac{3}{16}$

Order from least to greatest.

25. $\frac{2}{3}, \frac{4}{5}, \frac{6}{7}$ ______________________

26. $\frac{1}{5}, \frac{3}{11}, \frac{4}{9}$ ______________________

27. $\frac{3}{4}, \frac{7}{10}, \frac{2}{3}$ ______________________

28. $\frac{8}{9}, \frac{11}{12}, \frac{3}{4}$ ______________________

29. $\frac{2}{7}, \frac{1}{5}, \frac{3}{10}$ ______________________

30. $\frac{1}{10}, \frac{3}{15}, \frac{4}{45}$ ______________________

Use a calculator to change each fraction to a decimal rounded to the nearest thousandth. Then compare.

31. $\frac{42}{89}, \frac{96}{112}$ ______________________

32. $\frac{68}{310}, \frac{59}{289}$ ______________________

33. $\frac{92}{140}, \frac{86}{130}$ ______________________

34. $\frac{123}{456}, \frac{208}{802}$ ______________________

35. $\frac{76}{218}, \frac{159}{551}$ ______________________

36. $\frac{482}{999}, \frac{651}{1000}$ ______________________

*Use with Lesson 2-8, text pages 42–43.

Adding and Subtracting Fractions and Mixed Numbers*

Name ______________________

Date ______________________

Write the sum in simplest form.

1. $\frac{7}{10} + \frac{4}{5} =$ ______

2. $\frac{3}{4} + \frac{3}{5} =$ ______

3. $\frac{4}{11} + \frac{1}{3} =$ ______

4. $\frac{2}{5} + \frac{1}{7} =$ ______

5. $\frac{4}{5} + \frac{3}{8}$

6. $\frac{5}{9} + \frac{7}{12}$

7. $\frac{5}{8} + \frac{5}{6}$

8. $\frac{3}{7} + \frac{4}{5}$

9. $7\frac{3}{10} + 9\frac{7}{10}$

10. $2\frac{3}{11} + 7\frac{5}{11}$

11. $4\frac{3}{8} + 4\frac{1}{4}$

12. $8\frac{2}{3} + 3\frac{4}{9}$

13. $10\frac{5}{9} + 6\frac{3}{4}$

14. $2\frac{2}{7} + 8\frac{1}{2}$

15. $12\frac{5}{6} + 6\frac{4}{5}$

16. $2\frac{1}{3} + 5\frac{4}{7}$

Write the difference in simplest form.

17. $\frac{5}{12} - \frac{1}{12} =$ ______

18. $\frac{6}{7} - \frac{3}{7} =$ ______

19. $\frac{7}{8} - \frac{5}{8} =$ ______

20. $\frac{7}{9} - \frac{4}{9} =$ ______

21. $15\frac{5}{8} - 3\frac{1}{8}$

22. $4\frac{2}{9} - 1\frac{2}{9}$

23. $16\frac{7}{8} - 4\frac{3}{4}$

24. $14\frac{1}{3} - 8\frac{5}{6}$

25. $17\frac{5}{7} - 5\frac{2}{3}$

26. $8\frac{1}{4} - 2\frac{1}{3}$

27. $10\frac{2}{9} - 5\frac{5}{6}$

28. $11\frac{3}{4} - 9\frac{1}{2}$

Solve.

29. Three boxes weigh $6\frac{1}{4}$ lb, $4\frac{3}{8}$ lb, and $5\frac{3}{16}$ lb. How much do they weigh together? ______________________

30. The Gray family used $140\frac{5}{6}$ gallons of heating oil in January and $145\frac{3}{4}$ gallons in February. How many more gallons of oil did they use in February? ______________________

31. A designer has a piece of lace that measures $\frac{5}{8}$ yard. He used a $\frac{1}{3}$-yd piece to trim a collar. How much lace does he have left? ______________________

*Use with Lessons 2-9, 2-10, text pages 44–47.

Multiplying Fractions and Mixed Numbers*

Name ____________________

Date ____________________

Multiply. Cancel where possible.

1. $\frac{2}{3} \times \frac{4}{5} =$ ________
2. $\frac{7}{8} \times \frac{2}{3} =$ ________
3. $\frac{3}{5} \times \frac{1}{7} =$ ________
4. $\frac{5}{6} \times \frac{3}{4} =$ ________
5. $5 \times \frac{4}{15} =$ ________
6. $\frac{2}{3} \times 9 =$ ________
7. $\frac{7}{8} \times 14 =$ ________
8. $8 \times \frac{5}{6} =$ ________
9. $1\frac{3}{4} \times 2\frac{1}{3} =$ ________
10. $6\frac{1}{8} \times 1\frac{1}{7} =$ ________
11. $8\frac{1}{3} \times 2\frac{4}{5} =$ ________
12. $6\frac{1}{7} \times 3\frac{1}{2} =$ ________
13. $6\frac{1}{3} \times 4\frac{1}{5} =$ ________
14. $7\frac{1}{3} \times 9 =$ ________
15. $8 \times 4\frac{1}{2} =$ ________

Compute.

16. $(2\frac{1}{3} + 1\frac{5}{6}) \times \frac{4}{5} =$ ________
17. $(9\frac{1}{2} - 3\frac{3}{4}) \times \frac{8}{9} =$ ________
18. $(4\frac{3}{4} + 2\frac{3}{8}) \times 16 =$ ________
19. $(5\frac{3}{4} + 2\frac{1}{3}) \times \frac{4}{5} =$ ________
20. $(6\frac{2}{3} - \frac{4}{5}) \times \frac{1}{8} =$ ________
21. $14 \times (5\frac{1}{3} + \frac{4}{7}) =$ ________
22. $6\frac{1}{3} \times 4\frac{1}{5} \times \frac{5}{6} =$ ________
23. $7\frac{1}{7} \times 4\frac{2}{3} \times \frac{9}{10} =$ ________

Solve.

24. Twenty-four students went to the museum. One sixth of these saw the biology exhibit. How many saw the biology exhibit? ________
25. How many yards of fabric are needed to make 25 dolls if one doll requires $\frac{4}{5}$ yd of fabric? ________
26. Chris used $\frac{3}{5}$ of a $12\frac{1}{2}$-ft length of electrical wire. How much wire is left? ________

***Use with Lesson 2-11, text pages 48–49.**

Dividing Fractions, Complex Fractions, Mixed Numbers*

Name ____________________

Date ____________________

Write the reciprocal.

1. $\frac{3}{4}$ ______ 2. $\frac{2}{5}$ ______ 3. $\frac{1}{9}$ ______ 4. $\frac{2}{7}$ ______ 5. $\frac{4}{9}$ ______

6. 4 ______ 7. $2\frac{1}{6}$ ______ 8. $3\frac{2}{3}$ ______ 9. $\frac{7}{11}$ ______ 10. 9 ______

11. $8\frac{1}{3}$ ______ 12. $7\frac{2}{9}$ ______ 13. 18 ______ 14. $\frac{1}{12}$ ______ 15. $5\frac{2}{5}$ ______

Write the quotient in simplest form.

16. $\frac{3}{7} \div \frac{9}{10} =$ ______ 17. $\frac{4}{5} \div \frac{8}{9} =$ ______ 18. $\frac{3}{7} \div \frac{9}{14} =$ ______

19. $\frac{4}{9} \div \frac{2}{3} =$ ______ 20. $\frac{5}{12} \div \frac{5}{6} =$ ______ 21. $\frac{2}{9} \div \frac{11}{18} =$ ______

22. $18 \div \frac{6}{7} =$ ______ 23. $\frac{4}{5} \div 16 =$ ______ 24. $21 \div \frac{7}{8} =$ ______

25. $3\frac{1}{3} \div 1\frac{1}{2} =$ ______ 26. $5\frac{1}{8} \div \frac{1}{2} =$ ______ 27. $7\frac{3}{5} \div \frac{2}{15} =$ ______

28. $6\frac{2}{5} \div 2\frac{2}{3} =$ ______ 29. $5\frac{1}{4} \div 1\frac{2}{5} =$ ______ 30. $15 \div 8\frac{1}{3} =$ ______

Simplify.

31. $\dfrac{3\frac{2}{3}}{1\frac{1}{2}} =$ ______ 32. $\dfrac{2\frac{1}{8}}{\frac{1}{2}} =$ ______ 33. $\dfrac{2\frac{1}{5}}{\frac{2}{15}} =$ ______

34. $\dfrac{1\frac{1}{7}}{\frac{4}{21}} =$ ______ 35. $\dfrac{4\frac{3}{4}}{4\frac{1}{2}} =$ ______ 36. $\dfrac{8\frac{2}{3}}{1\frac{2}{9}} =$ ______

Solve.

37. Stephanie bikes $82\frac{1}{2}$ miles in $6\frac{3}{4}$ hours. How many miles does she travel in one hour? ______

*Use with Lesson 2-12, text pages 50–51.

Estimating Fractions*

Name ______________________

Date ______________________

Estimate each sum and difference. Use the estimation strategy that works best for you.

1. $5\frac{1}{15} + 6\frac{7}{30}$ ______ **2.** $3\frac{4}{7} + 9\frac{2}{11}$ ______ **3.** $4\frac{7}{10} + 5\frac{6}{35}$ ______ **4.** $11\frac{4}{5} + 9\frac{10}{11}$ ______

5. $8\frac{17}{18} - 4\frac{5}{8}$ ______ **6.** $7\frac{1}{7} - 5\frac{2}{5}$ ______ **7.** $14\frac{9}{10} - 9\frac{2}{31}$ ______ **8.** $16\frac{19}{20} - 7\frac{2}{5}$ ______

9. $\frac{7}{8} + 2\frac{5}{16} + \frac{2}{3} + 8\frac{5}{12} + 1\frac{3}{4}$ ______ **10.** $11\frac{4}{5} + 2\frac{8}{9} + 6\frac{2}{15} + 10\frac{3}{20} + \frac{5}{6}$ ______

Estimate each product.

11. $\frac{2}{3} \times 19$ ______ **12.** $\frac{5}{6} \times 31$ ______ **13.** $\frac{11}{12} \times 56$ ______ **14.** $\frac{9}{20} \times 41$ ______

15. $\frac{1}{6} \times 11$ ______ **16.** $\frac{1}{9} \times \frac{19}{20}$ ______ **17.** $\frac{7}{8} \times \frac{31}{40}$ ______ **18.** $\frac{1}{5} \times \frac{21}{23}$ ______

19. $2\frac{2}{3} \times \frac{1}{15}$ ______ **20.** $9\frac{1}{3} \times \frac{1}{13}$ ______ **21.** $\frac{1}{4} \times 3\frac{4}{7}$ ______ **22.** $\frac{7}{9} \times 2\frac{1}{5}$ ______

Solve mentally. (Remember: sometimes the *distributive property* can be used.)

23. $\frac{1}{2} \times 8\frac{2}{3}$ ______ **24.** $\frac{2}{5} \times 5\frac{1}{11}$ ______ **25.** $\frac{2}{7} \times 21\frac{2}{3}$ ______ **26.** $\frac{1}{8} \times 32\frac{3}{5}$ ______

27. $\frac{3}{4} \times 12\frac{1}{8}$ ______ **28.** $\frac{2}{3} \times 15\frac{4}{5}$ ______ **29.** $\frac{7}{9} \times 18\frac{3}{7}$ ______ **30.** $\frac{2}{11} \times 22\frac{1}{3}$ ______

31. $\frac{8}{9} \times 54\frac{4}{5}$ ______ **32.** $\frac{3}{16} \times 32\frac{1}{3}$ ______ **33.** $\frac{9}{10} \times 30\frac{2}{3}$ ______ **34.** $\frac{5}{8} \times 48\frac{1}{6}$ ______

Solve using estimation.

35. An Amazon caterpillar grows $\frac{4}{15}$ of an inch each week. If it keeps growing at this rate, how long will it grow in $3\frac{2}{3}$ years? ______________________

36. A desert caravan carried $8\frac{3}{4}$ gallons of drinking water. If $\frac{2}{3}$ of the water was used on a trip, about how much was left? ______________________

*Use with Lesson 2-13, text pages 52–53.

Order of Operations*

Name ______________________

Date ______________________

Simplify. $20 \div 5 \times 3 = 4 \times 3 = 12$

1. $14 + 3 - 7 =$ ______

2. $9 - 4 + 10 =$ ______

3. $7 + 3 - 2 =$ ______

4. $3 \times 4 \div 2 =$ ______

5. $36 \div 9 \times 5 =$ ______

6. $6 \times 4 \div 12 =$ ______

7. $2 \times 7 + 5 =$ ______

8. $4 \times 3 + 6 =$ ______

9. $9 - 3 \div 3 =$ ______

10. $42 \div 7 - 6 =$ ______

11. $3 \times 5 \div 5 + 1 =$ ______

12. $48 \div 3 \times 2 + 5 =$ ______

13. $[4(7 + 3)] - 5 =$ ______

14. $2[6(5 - 3)] =$ ______

15. $[3(5 + 4) - 3] =$ ______

16. $5[4 + 3] + 10 =$ ______

17. $9(7 - 5) \div 3 =$ ______

18. $5(20 - 12) \div 4 =$ ______

Simplify.

19. $[(5 \times 8) + 9] \div [(10 - 6) + 3] =$ ______

20. $[20 \div (30 - 26)] \times [(56 + 48) \div 26] =$ ______

21. $[2(6 \times 2) \div 6] + [(24 + 11) \div 5] =$ ______

22. $\{3[2 \times (37 - 12)] - 5\} - (324 \div 9) =$ ______

23. $[14 \times 7 - 3(4 \times 2) + 8] + 10 \div 2 =$ ______

24. $2(44) + [42 - 3(5 + 3)] - 21 =$ ______

25. $[2(45 \div 5)] - [36 \div (3 \times 3)] =$ ______

*Use with Lesson 3-1, text pages 64–65.

Mathematical or Algebraic Expressions*

Name ______________________

Date ______________________

Write a mathematical expression. Choose a letter for each variable.

1. 8 less than a number ______________________

2. a number divided by 3 ______________________

3. 5 times a number ______________________

4. 4 more than a number ______________________

5. a number less 9 ______________________

6. 15 divided by a number ______________________

7. 12 times a number ______________________

8. 2 less than a number ______________________

9. a number divided into 4 parts ______________________

10. 7 more than twice a number ______________________

11. 1 more than a number divided by 3 ______________________

12. 30 less than a number ______________________

13. 9 more than 7 times a number ______________________

14. 5 added to a number ______________________

15. the product of a number and 5 ______________________

16. 7 less than a number ______________________

17. half of a number ______________________

18. 6 times a number ______________________

19. 4 less than 5 times a number ______________________

20. a number divided by 3 increased by 2 ______________________

Write a word phrase for each.

21. $7n + 9$ ______________________

22. $4a + 5$ ______________________

***Use with Lesson 3-2, text pages 66–67.**

Mathematical or Algebraic Sentences*

Name ______________________

Date ______________________

Write a mathematical or algebraic sentence. Identify each as an equation or an inequality.

1. The sum of a number and 4 is 10. ______________________
2. One fifth of a number is 20. ______________________
3. 5 less than a quarter of a number is greater than 1. ______________________
4. The difference between a number and 3 is not 6. ______________________
5. The product of a number and 5 is less than 10. ______________________
6. The quotient of a number and 10 is 5. ______________________
7. 3 increased by a number is less than 6. ______________________
8. 2 less than 3 times a number is less than or equal to 11. ______________________
9. 5 less than half a number is greater than or equal to 6. ______________________

Write a mathematical sentence to describe each problem situation.

10. The perimeter of a square is 24 cm. Find the length of one side. ______________________
11. Jennifer is 6 years older than her sister Amanda. If Jennifer is 14, how old is Amanda? ______________________

Write an English sentence for each.

12. $2y - 1 = 1$ ______________________
13. $\frac{z}{3} \geq 1$ ______________________
14. $a + 2 \neq 7$ ______________________
15. $\frac{x}{4} + 2 \leq 5$ ______________________

____

*Use with Lesson 3-4, text pages 70–71.

Evaluating Expressions and Sentences*

Name ____________________

Date ____________________

Find the value when $a = 10$, $b = 25$, $c = 32$.

1. $4a$ ________
2. $b - 7$ ________
3. $a + 13$ ________
4. $a \div 2$ ________
5. $12 + c$ ________
6. $b - 17$ ________
7. $b \div 5$ ________
8. $4b$ ________
9. $c \div 16$ ________
10. $10a$ ________
11. $c \div 2$ ________
12. $14.2a$ ________
13. $c \div 4$ ________
14. $b + 5$ ________
15. $8a$ ________
16. $7 + c$ ________
17. $100 + b$ ________
18. $c - 21$ ________

Evaluate each expression when $r = 90$, $s = 50$, $t = 200$, $v = 3$.

19. $\frac{s}{2} + 6$ ________
20. $(r + 10) \div 5$ ________
21. $\frac{t}{4} \div 2$ ________
22. $\frac{t}{s} + 30$ ________
23. $(r - s) \div 8$ ________
24. $(70 - s) + r$ ________
25. $\frac{r}{v} + 9$ ________
26. $10v + r$ ________
27. $5s - t$ ________
28. $2r - \frac{3s}{v}$ ________
29. $(r + 7) - (s + 9)$ ________
30. $vt - vs$ ________

Evaluate each expression when $m = 40$, $n = 120$, $w = 100$, $x = 75$.

31. $\frac{m}{2} - (n - w)$ ________
32. $2(m + x) - w$ ________
33. $\frac{4n}{m} + w$ ________
34. $(2n - w) + 2x$ ________
35. $2(m + n) - w$ ________
36. $2m \div \frac{3w}{x}$ ________
37. $\frac{w}{4} + 3x$ ________
38. $2x - (n \div m)$ ________

*Use with Lesson 3-3, text pages 68–69.

Addition and Subtraction Equations*

Name ______________________

Date ______________________

Solve. $n + 6 = 9 \quad n + 6 - 6 = 9 - 6 \quad n = 3$

1. $n + 6 = 9$ $\quad n =$ ______
2. $r + 8 = 14$ $\quad r =$ ______
3. $v + 7 = 21$ $\quad v =$ ______
4. $t + 2 = 10$ $\quad t =$ ______
5. $x + 3 = 10$ $\quad x =$ ______
6. $b + 4 = 20$ $\quad b =$ ______
7. $a + 15 = 25$ $\quad a =$ ______
8. $30 + c = 46$ $\quad c =$ ______
9. $b + 12 = 21$ $\quad b =$ ______
10. $42 = d + 11$ $\quad d =$ ______
11. $19 = s + 6$ $\quad s =$ ______
12. $9 + r = 14$ $\quad r =$ ______
13. $n - 2 = 6$ $\quad n =$ ______
14. $a - 3 = 12$ $\quad a =$ ______
15. $r - 3 = 4$ $\quad r =$ ______
16. $b - 8 = 16$ $\quad b =$ ______
17. $s - 7 = 13$ $\quad s =$ ______
18. $t - 21 = 30$ $\quad t =$ ______
19. $b - 9 = 18$ $\quad b =$ ______
20. $c - 17 = 41$ $\quad c =$ ______
21. $d - 8 = 42$ $\quad d =$ ______
22. $f - 9 = 6$ $\quad f =$ ______
23. $h - 6 = 94$ $\quad h =$ ______
24. $k - 12 = 7$ $\quad k =$ ______
25. $a + 26 = 38$ $\quad a =$ ______
26. $b - 40 = 20$ $\quad b =$ ______
27. $c - 3 = 81$ $\quad c =$ ______
28. $d + 14 = 30$ $\quad d =$ ______
29. $21 + r = 38$ $\quad r =$ ______
30. $50 = a + 7$ $\quad a =$ ______
31. $t - 16 = 75$ $\quad t =$ ______
32. $s + 16 = 80$ $\quad s =$ ______
33. $r - 9 = 51$ $\quad r =$ ______
34. $d - 32 = 140$ $\quad d =$ ______
35. $f + 6 = 21$ $\quad f =$ ______
36. $c - 7 = 48$ $\quad c =$ ______

*Use with Lesson 3-6, text pages 74–75.

Multiplication and Division Equations*

Name ______________________

Date ______________________

Solve.

1. $7x = 42$

$x =$ ______

2. $9r = 36$

$r =$ ______

3. $4n = 80$

$n =$ ______

4. $6t = 24$

$t =$ ______

5. $5y = 100$

$y =$ ______

6. $3m = 96$

$m =$ ______

7. $2a = 144$

$a =$ ______

8. $9b = 288$

$b =$ ______

9. $6c = 90$

$c =$ ______

10. $600 = 12n$

$n =$ ______

11. $7r = 112$

$r =$ ______

12. $228 = 4y$

$y =$ ______

13. $\frac{b}{6} = 12$

$b =$ ______

14. $\frac{a}{9} = 10$

$a =$ ______

15. $\frac{c}{4} = 21$

$c =$ ______

16. $\frac{a}{3} = 25$

$a =$ ______

17. $\frac{n}{7} = 14$

$n =$ ______

18. $\frac{x}{5} = 32$

$x =$ ______

19. $\frac{r}{27} = 3$

$r =$ ______

20. $\frac{s}{10} = 13$

$s =$ ______

21. $46 = \frac{t}{4}$

$t =$ ______

22. $\frac{c}{3} = 27$

$c =$ ______

23. $25 = \frac{a}{5}$

$a =$ ______

24. $\frac{n}{14} = 14$

$n =$ ______

25. $12a = 48$

$a =$ ______

26. $\frac{a}{8} = 6$

$a =$ ______

27. $5r = 300$

$r =$ ______

28. $\frac{n}{12} = 10$

$n =$ ______

29. $7c = 455$

$c =$ ______

30. $\frac{d}{6} = 30$

$d =$ ______

31. $\frac{j}{3} = 12$

$j =$ ______

32. $9m = 405$

$m =$ ______

33. $\frac{d}{4} = 48$

$d =$ ______

***Use with Lesson 3-7, text pages 76–77.**

Equations*
(With More Than One Operation)

Name ______________________

Date ______________________

Solve. $3x + 1 = 13$ $\quad 3x + 1 - 1 = 13 - 1$ $\quad 3x = 12$ $\quad x = 4$

1. $4a + 2 = 30$ $\quad a =$ ______
2. $\frac{r}{7} + 5 = 10$ $\quad r =$ ______
3. $\frac{b}{6} - 6 = 6$ $\quad b =$ ______
4. $7c - 6 = 50$ $\quad c =$ ______
5. $3n + 10 = 22$ $\quad n =$ ______
6. $5x - 12 = 28$ $\quad x =$ ______
7. $\frac{n}{9} + 4 = 12$ $\quad n =$ ______
8. $9r - 14 = 67$ $\quad r =$ ______
9. $2a + 53 = 67$ $\quad a =$ ______
10. $\frac{s}{4} + 5 = 14$ $\quad s =$ ______
11. $10a - 7 = 63$ $\quad a =$ ______
12. $\frac{x}{11} + 6 = 10$ $\quad x =$ ______
13. $\frac{4n + 4}{2} = 10$ $\quad n =$ ______
14. $7n - 54 = 16$ $\quad n =$ ______
15. $\frac{a}{5} + 7 = 20$ $\quad a =$ ______
16. $\frac{n}{5} - 7 = 18$ $\quad n =$ ______
17. $6x + 11 = 47$ $\quad x =$ ______
18. $6b - 29 = 1$ $\quad b =$ ______
19. $2a + 4 = 14$ $\quad a =$ ______
20. $8x - 2 = 38$ $\quad x =$ ______
21. $\frac{n}{4} + 5 = 9$ $\quad n =$ ______
22. $\frac{a}{14} + 10 = 12$ $\quad a =$ ______
23. $10x - 13 = 37$ $\quad x =$ ______
24. $\frac{r}{12} + 21 = 24$ $\quad r =$ ______

Solve.

25. The school science lab has 5 times as many microscopes now as it did last year. If there are now 75 microscopes, how many were there last year? ______________________

*Use with Lesson 3-8, text pages 78–79.

Equations*
(With One or More Operations)

Name ______________________

Date ______________________

Solve and check.

1. $a + 19 = 46$

$a =$ ______

2. $b - 38 = 10$

$b =$ ______

3. $c - 4 = 96$

$c =$ ______

4. $d + 15 = 130$

$d =$ ______

5. $7n = 56$

$n =$ ______

6. $\frac{r}{6} = 42$

$r =$ ______

7. $\frac{2a}{3} + 2 = 8$

$a =$ ______

8. $5(n + 3) - 5 = 40$

$n =$ ______

9. $9 + \frac{a}{5} = 10$

$a =$ ______

10. $42 + 3x = 54$

$x =$ ______

11. $6c - 29 = 7$

$c =$ ______

12. $\frac{5b}{2} + 7 = 17$

$b =$ ______

13. $8n - 14 = 26$

$n =$ ______

14. $\frac{7x + 9}{5} - 12 = 8$

$x =$ ______

15. $6(n + 8) = 96$

$n =$ ______

Solve.

16. On the first night of the science exhibit, 1427 people viewed projects. 369 fewer people came the second night. How many people came then? ______________________

17. There are 426 entries in the science fair. If this is 53 more than the last fair, how many entries were there in the earlier fair? ______________________

18. Mark read a book of 189 pages. If this is three times the number of pages in Adam's book, how many pages are there in Adam's book? ______________________

19. Thirty-five students belong to the science club. This is $\frac{1}{4}$ the number in the 5 sections of the eighth grade. How many students are in the eighth grade? How many in each section if each section has the same number of students? ______________________

*Use with Lesson 3-8, text pages 78–79.

Equations with Grouping Symbols*

Name ______________________

Date ______________________

Solve each equation by using the distributive property.

$$\begin{aligned} 5(n - 3) &= 40 \\ 5n - 15 &= 40 \\ 5n - 15 + 15 &= 40 + 15 \\ \frac{5n}{5} &= \frac{55}{5} \\ n &= 11 \end{aligned}$$

1. $4(a + 3) = 32$ ________

2. $7(r - 7) = 49$ ________

3. $5(y + 2) = 15$ ________

4. $14(n - 9) = 42$ ________

5. $8(s + 6) = 64$ ________

6. $13(c + 8) = 143$ ________

Solve each equation by using inverse operations.

7. $2(x + 8) = 48$ ________

8. $12(h + 6) = 84$ ________

9. $6(s - 8) = 30$ ________

10. $(c + 5)25 = 600$ ________

11. $(m - 9)8 = 64$ ________

12. $9(k + 6) = 108$ ________

Solve each equation.

13. $\frac{a - 6}{8} + 4 = 16$ ________

14. $\frac{b + 9}{3} + 5 = 28$ ________

15. $\frac{n + 4}{5} + 12 = 16$ ________

16. $\frac{3x - 2}{4} + 5 = 15$ ________

17. $\frac{n + 7}{2} + 2 = 11$ ________

18. $\frac{8y - 2}{6} - 8 = 21$ ________

19. $\frac{5d + 3}{4} - 1 = 18$ ________

20. $\frac{t + 8}{2} - 6 = 3$ ________

21. $\frac{d - 5}{6} - 3 = 10$ ________

22. $\frac{8s + 14}{4} - 4 = 6$ ________

23. $\frac{3r + 5}{2} + 2 = 12$ ________

24. $\frac{7y + 2}{6} - 3 = 22$ ________

Solve. Write an equation.

25. The product of 8 and the difference between a number and 9 is 96. What is the number? ______________________

26. When 12 is added to half the sum of a number and 8 the result is 40. What is the number? ______________________

27. The difference between a number and 5, when multiplied by 3 is 36. What is the number? ______________________

*Use with Lesson 3-9, text pages 80–81.

Solutions for Inequalities*

Name ______________________

Date ______________________

Complete the chart.

	Inequality	Replacement Set	Solution Set
1.	$r > 3$	{0, 1, 2, 3, 4}	
2.	$s \leq 4$	{0, 1, 2, 3, 4}	
3.	$9 < x$	{5, 6, 7, . . . , 25}	
4.	$11 \neq r$	{5, 6, 7, 8, 9}	
5.	$b \geq 21$	{20, 25, 30, . . . , 50}	
6.	$c \leq 7$	{100, 99, 98, . . . , 0}	
7.	$19 \neq a$	{17, 19, 21, . . . , 33}	
8.	$26 \leq c$	{100, 200, 300}	
9.	$3 \geq d$	{0, 1, 2, . . . , 10}	
10.	$x \neq 13$	{12, 14, 16, 18, 20}	
11.	$c \leq 29$	{50, 40, 30, . . . , 10}	
12.	$y < 6$	{0, 1, 2, . . .}	

Write <, ≤, >, ≥, or = to form an inequality that matches the given solution set, S.
The replacement set is {0, 1, 2, 3, 4, 5, 6, 7, 8}

13. n _____ 4 S: {0, 1, 2, 3}

14. r _____ 4 S: {5, 6, 7, 8}

15. x _____ 4 S: {4, 5, 6, 7, 8}

16. a _____ 4 S: {0, 1, 2, 3, 4}

17. s _____ 4 S: {0, 1, 2, 3, 5, 6, 7, 8}

18. a _____ 7 S: {0, 1, 2, 3, 4, 5, 6}

List the members of each set.

19. The set of whole numbers between 20 and 30 ______________________

20. The set of odd numbers < 13 ______________________

21. The set of multiples of 3 less than 21 ______________________

22. The set of multiples of 10 ______________________

*Use with Lesson 3-10, text pages 82–83.

Using Formulas*

Name ______________________

Date ______________________

What kind of problem does each formula solve?

1. $A = s^2$ ____________
2. $A = \pi r^2$ ____________
3. $C = \pi d$ ____________
4. $P = 4s$ ____________
5. $A = \frac{1}{2}bh$ ____________
6. $V = \ell wh$ ____________

Solve for the missing variable.

7. $P = 4s$ when $P = 64$ ____________
8. $C = \pi d$ when $d = 42$ ____________
9. $V = \ell wh$ when $V = 75$ cubic feet, $w = 3'$, and $h = 5'$ ____________
10. $A = \frac{1}{2}bh$ when $b = 6'$, and $h = 21'$ ____________
11. $A = s^2$ when $A = 324$ square yards ____________
12. $A = \pi r^2$ when $r = 9$ ____________

Write a formula for each.

13. Rate of Tax = Tax divided by Marked Price ____________
14. Total Sales = Commission divided by Rate of Commission ____________
15. Discount = List Price times Rate of Discount ____________
16. Commission = Total Sales times Rate of Commission ____________

Solve, using formulas.

17. The outside rim of a circular pool measures 68 ft. What is the radius of the pool? ____________
18. A square board is 0.6 m long. Find its area. ____________
19. An aquarium is 20 feet long, 4 feet wide and 3 feet deep. How many cubic feet of water will it hold? ____________
20. How long will it take $3200 to earn $400 at 6% a year? ____________

*Use with Lesson 3-11, text pages 84–85.

Integers*

Name ____________________

Date ____________________

Write the opposite of each integer.

1. $^{+}5$ ______ **2.** $^{-}6$ ______ **3.** $^{+}4$ ______ **4.** $^{+}102$ ______ **5.** $^{+}17$ ______

Compare. Write <, =, or >.

6. $^{+}3$ ______ $^{-}12$ **7.** $^{-}41$ ______ $^{-}50$ **8.** $^{+}1$ ______ $^{-}20$

9. $^{-}3$ ______ 0 **10.** $^{-}100$ ______ $^{+}1$ **11.** $^{-}16$ ______ $^{+}61$

12. $^{+}2$ ______ $^{-}5$ **13.** $^{+}21$ ______ $^{-}12$ **14.** $^{+}30$ ______ $^{+}3$

Arrange in order from least to greatest.

15. $^{+}5$, $^{+}2$, $^{-}20$, $^{-}15$, $^{+}20$ ____________________

16. $^{-}10$, $^{+}6$, $^{+}16$, $^{+}10$, $^{-}6$ ____________________

Arrange in order from greatest to least.

17. $^{-}19$, 0, $^{+}15$, $^{+}5$, $^{-}5$ ____________________

18. $^{-}10$, $^{+}20$, $^{-}30$, 0, $^{-}15$ ____________________

Write an integer for each expression.

19. a deduction of $31.00 ____________

20. a loss of 18 points ____________

21. 500 meters above sea level ____________

22. a 20° rise in temperature ____________

Solve.

23. Last Monday the temperature was $^{-}12$°C. It rose 9°C this Monday. What is the temperature this Monday? ____________

24. The temperature is 28°F but the wind chill makes it feel 18° colder. What is the wind chill temperature? ____________

25. How many integers are there between $^{-}13$ and $^{+}13$? Between $^{+}310$ and $^{-}300$? Between $^{-}15$ and 0? ____________

*Use with Lesson 3-12, text pages 86–87.

Operations with Integers: Addition and Subtraction*

Name ____________________

Date ____________________

Add. $^{+}7 - {}^{+}10 = {}^{-}3$ $^{+}10 - {}^{+}7 = {}^{+}3$

1. $^{+}7 + {}^{+}2$ ______ **2.** $^{-}6 + {}^{+}2$ ______ **3.** $^{-}1 + {}^{+}8$ ______ **4.** $^{+}9 + {}^{+}8$ ______

5. $^{-}5 + 0$ ______ **6.** $^{+}3 + {}^{-}3$ ______ **7.** $0 + {}^{+}10$ ______ **8.** $^{-}8 + {}^{-}3$ ______

9. $^{-}6 + {}^{+}6$ ______ **10.** $^{-}13 + 0$ ______ **11.** $^{-}1 + {}^{-}1$ ______ **12.** $^{+}7 + {}^{-}6$ ______

13. $^{+}7 + 0$ ______ **14.** $^{-}1 + {}^{+}6$ ______ **15.** $0 + {}^{-}2$ ______ **16.** $^{+}3 + {}^{-}10$ ______

Subtract.

17. $^{+}3 - {}^{-}11$ ______ **18.** $^{+}12 - {}^{-}12$ ______ **19.** $^{-}8 - {}^{+}17$ ______ **20.** $^{-}14 - {}^{-}24$ ______

21. $^{+}30 - {}^{-}7$ ______ **22.** $^{+}6 - {}^{+}18$ ______ **23.** $0 - {}^{-}20$ ______ **24.** $0 - {}^{+}32$ ______

25. $^{-}13 - {}^{-}13$ ______ **26.** $0 - {}^{-}1$ ______ **27.** $^{-}28 - {}^{+}5$ ______ **28.** $^{+}8 - {}^{-}16$ ______

29. $0 - {}^{+}22$ ______ **30.** $^{+}15 - {}^{+}41$ ______ **31.** $^{-}9 - {}^{+}9$ ______ **32.** $^{+}2 - {}^{-}6$ ______

Solve each addition or subtraction equation.

33. $^{-}4 + b = {}^{-}6$ ______ **34.** $^{-}5 + d = {}^{+}4$ ______ **35.** $x - {}^{+}3 = {}^{+}8$ ______

36. $^{+}5 + c = {}^{+}13$ ______ **37.** $c - {}^{+}3 = {}^{+}12$ ______ **38.** $y - {}^{-}4 = {}^{-}7$ ______

39. $a - {}^{-}8 = {}^{+}7$ ______ **40.** $h - {}^{+}5 = {}^{+}9$ ______ **41.** $x - {}^{-}10 = {}^{+}7$ ______

42. $^{-}6 + x = {}^{-}8$ ______ **43.** $c - {}^{-}8 = {}^{+}5$ ______ **44.** $a + {}^{+}7 = {}^{-}3$ ______

Solve.

45. Mike's checking account has a balance of $^{-}\$44$. If he deposits \$128, what will be the new balance? ____________

46. The temperature this morning was 2°F. It rose to 21° and then dropped by 15°. What was the final temperature? ____________

47. What number would you subtract from $^{-}31$ to give you a $^{+}19$? ____________

*Use with Lesson 3-13, text pages 88–89.

Operations with Integers: Multiplication and Division*

Name ______________________

Date ______________________

Multiply. $^{+}5 \times {}^{-}8 = {}^{-}40$

1. $^{+}6 \times {}^{+}12$ ______
2. $^{-}3 \times {}^{+}13$ ______
3. $^{-}9 \times 0$ ______
4. $^{+}8 \times {}^{-}5$ ______
5. $^{-}2 \times {}^{-}2$ ______
6. $^{+}9 \times {}^{-}8$ ______
7. $^{-}4 \times {}^{-}7$ ______
8. $^{-}2 \times {}^{-}14$ ______
9. $^{-}18 \times {}^{-}5$ ______
10. $^{-}22 \times {}^{+}7$ ______
11. $^{+}31 \times {}^{-}3$ ______
12. $^{-}100 \times {}^{+}100$ ______

Divide.

13. $\frac{^{-}40}{^{-}4}$ ______
14. $\frac{^{-}144}{^{-}12}$ ______
15. $\frac{^{-}45}{^{+}15}$ ______
16. $\frac{^{+}54}{^{-}6}$ ______
17. $\frac{^{+}63}{^{-}7}$ ______
18. $\frac{^{-}300}{^{+}20}$ ______
19. $\frac{^{+}225}{^{-}25}$ ______
20. $\frac{^{-}72}{^{+}12}$ ______
21. $\frac{^{-}119}{^{+}7}$ ______
22. $\frac{^{+}315}{^{-}5}$ ______
23. $\frac{^{-}135}{^{-}9}$ ______
24. $\frac{^{-}168}{^{+}14}$ ______

Solve each multiplication or division equation.

25. $^{+}12t = {}^{-}144$ ______
26. $^{-}9d = {}^{+}63$ ______
27. $^{+}20a = 0$ ______
28. $\frac{c}{^{+}12} = 0$ ______
29. $\frac{s}{^{-}6} = {}^{-}6$ ______
30. $\frac{n}{^{+}9} = {}^{-}14$ ______

Solve.

31. A number is multiplied by $^{+}2$, then multiplied by $^{-}4$. The result is $^{+}48$. What is the number? ______

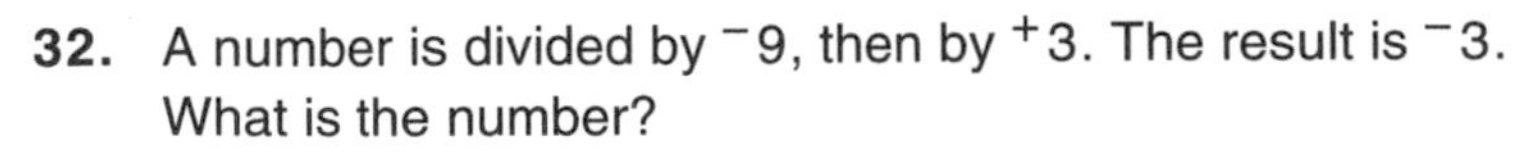

32. A number is divided by $^{-}9$, then by $^{+}3$. The result is $^{-}3$. What is the number? ______

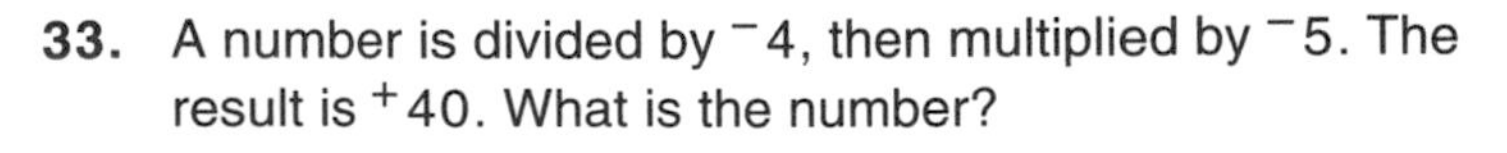

33. A number is divided by $^{-}4$, then multiplied by $^{-}5$. The result is $^{+}40$. What is the number? ______

34. A number is multiplied by $^{-}8$, then divided by $^{-}4$. The result is $^{+}4$. What is the number? ______

*Use with Lesson 3-13, text pages 88–89.

Equations with Integers*

Name ______________________

Date ______________________

Solve by writing and solving an equation.

1. Lisa has 14 coins. She has some quarters, 5 nickels, and one dime more than she has nickels. How many of each type of coin does she have? How much money does she have? ______________________

2. The sum of a number and 98 is 150. What is the number? ______________________

3. The sum of eight and four times a number is forty. What is the number? ______________________

4. The difference between 87 and a number is 31. Find the number. ______________________

5. Six times a number increased by five is fifty-three. What is the number? ______________________

6. Ten more than half a number is forty-eight. What is the number? ______________________

7. A race car is traveling at 120 mph. At this rate, how long will a race of 1500 miles take? ______________________

8. One number is nine times another. The greater number is 4950. What is the lesser number? ______________________

9. The length of a rectangle is four times its width. The length is 34 cm. Find the area. ______________________

10. The Tigers scored 16 more points in the second half than in the first half. They scored 86 points in all. How many points did they score in each half? ______________________

***Use with Lesson 3-15, text pages 92–93.**

Rational Numbers*

Name ______________________

Date ______________________

Write the opposite of each rational number.

1. $^{-}0.2$ ______ **2.** $^{+}6.11$ ______ **3.** $^{-}\frac{5}{3}$ ______ **4.** $^{-}\frac{6}{4}$ ______ **5.** $^{-}8.1$ ______

6. $^{-}212$ ______ **7.** 0 ______ **8.** $^{-}3.0\overline{1}$ ______ **9.** $^{-}\frac{9}{2}$ ______ **10.** $^{+}\frac{8}{4}$ ______

Compare. Write $<$, $=$, or $>$.

11. $^{+}5$ ____ $^{-}6$ **12.** $^{-}4$ ____ $^{+}11$ **13.** 0 ____ $^{+}7.2$ **14.** $^{+}0.9$ ____ $^{-}\frac{3}{4}$

15. $^{+}3.9$ ____ $^{+}1\frac{1}{2}$ **16.** $^{-}3.6$ ____ 0 **17.** $^{-}\frac{1}{4}$ ____ 0.1 **18.** $^{+}10$ ____ $^{-}10.2$

19. $^{-}0.\overline{3}$ ____ $^{-}\frac{1}{8}$ **20.** $^{-}\frac{2}{3}$ ____ $^{+}0.7\overline{5}$ **21.** $^{-}\frac{4}{5}$ ____ $^{-}0.6$ **22.** $^{-}0.\overline{1}$ ____ $^{-}\frac{1}{10}$

Write each set of numbers in order from least to greatest.

23. $^{+}2\frac{1}{2}$, $^{-}\frac{7}{2}$, $^{-}0.1$, $^{+}3\frac{2}{3}$, $^{-}3$ ______________________________

24. $^{-}\frac{11}{3}$, $^{-}5$, $^{-}2\frac{1}{4}$, $^{+}2.25$, $^{-}0.1$ ______________________________

25. $^{+}3.10$, $^{-}0.31$, $^{-}1.3$, $^{+}0.31$, $^{-}\frac{1}{3}$ ______________________________

Write a rational number for each expression.

26. a deposit of \$141.15 ______________________________

27. 5.03 km underwater ______________________________

28. 20.4° below zero ______________________________

29. down $4\frac{2}{3}$ points ______________________________

30. a gain of \$16.25 ______________________________

***Use with Lesson 4-1, text pages 100–101.**

Properties of Rational Numbers*

Name ______________________

Date ______________________

Name the property.

1. $^{-}7 \times {}^{+}4 = {}^{+}4 \times {}^{-}7$

   ____

2. $^{+}4 \times (^{-}1 + {}^{+}3) = (^{+}4 \times {}^{-}1) + (^{+}4 \times {}^{+}3)$

   ____

3. $^{+}6 + {}^{-}6 = 0$

   ____

4. $(^{-}4 + {}^{+}5) + {}^{-}2 = {}^{-}4 + (^{+}5 + {}^{-}2)$

   ____

5. $^{+}6 + {}^{-}3 = {}^{-}3 + {}^{+}6$

   ____

6. $^{-}8 \times -\frac{1}{8} = 1$

   ____

7. $^{+}9 + 0 = {}^{+}9$

   ____

8. $^{-}3 \times 1 = {}^{-}3$

   ____

Simplify. Use the number properties.

9. $^{-}12 + (^{+}3 + {}^{-}3) =$ ______________

10. $^{+}5 \times (^{-}6 + {}^{+}7) =$ ______________

11. $+\frac{24}{5} \times \frac{5}{(^{-}8 \times {}^{-}3)} =$ ______________

12. $(^{+}21 \times {}^{-}14) \times {}^{-}10 =$ ______________

13. $(^{-}9 + {}^{-}6) - {}^{+}4 =$ ______________

14. $(^{+}19 + {}^{+}48 + {}^{-}67) =$ ______________

15. $(^{+}6 + {}^{-}6) \times (^{-}4 + {}^{+}4) =$ ______________

16. $^{-}246 \times {}^{+}308 \times -\frac{1}{246} =$ ______________

17. $^{-}48 + {}^{-}92 + {}^{+}48 =$ ______________

18. $(^{-}2 \times {}^{+}4) + (^{-}2 \times {}^{-}8) =$ ______________

19. $(^{+}17 + {}^{-}24) - {}^{-}16 =$ ______________

20. $+\frac{1}{3} \times -\frac{3}{4} \times -\frac{8}{9} =$ ______________

21. $(^{-}3.4 - {}^{-}2.1) + {}^{-}2.1 =$ ______________

22. $^{+}6.2 \times (^{-}7 + {}^{+}1) =$ ______________

*Use with Lesson 4-2, text pages 102–103.

Adding Rational Numbers*

Name ______________________

Date ______________________

To add with *like* signs:
- Find the sum of the numbers.
- Use the sign of the addends.

To add with *unlike* signs:
- Find the difference of the numbers.
- Use the sign of the addend farther from zero (one with greater absolute value).

Find the sum.

1. $^{+}2.5 + {}^{+}5.6 =$ ________ **2.** $^{+}6.1 + {}^{+}8.15 =$ ________ **3.** $^{+}0.3 + {}^{+}7.2 =$ ________

4. $^{+}7.5 + {}^{+}7.8 =$ ________ **5.** $^{+}7.04 + {}^{+}8.7 =$ ________ **6.** $^{-}0.43 + {}^{-}0.56 =$ ________

7. $^{+}5.6 + {}^{-}0.01 =$ ________ **8.** $^{+}2.2 + {}^{-}6.6 =$ ________ **9.** $^{-}0.15 + {}^{-}1.8 =$ ________

10. $^{+}7.8 + {}^{-}2.1 =$ ________ **11.** $^{-}7.5 + {}^{+}2.5 =$ ________ **12.** $^{+}0.15 + {}^{-}0.61 =$ ________

13. $^{-}8.7 + {}^{+}7.8 =$ ________ **14.** $^{+}0.16 + {}^{-}0.25 =$ ________ **15.** $^{+}0.28 + {}^{-}0.35 =$ ________

16. $^{-}10.1 + {}^{+}96.1 =$ ________ **17.** $^{+}3.4 + {}^{-}0.8 =$ ________ **18.** $^{-}0.6 + {}^{+}1 =$ ________

19. $^{-}9.7 + {}^{+}8.9 =$ ________ **20.** $^{+}1.9 + {}^{-}4.3 =$ ________ **21.** $\frac{-5}{8} + \frac{+2}{3} =$ ________

22. $^{+}3\frac{1}{2} + {}^{-}3\frac{1}{2} =$ ________ **23.** $^{-}7.81 + {}^{+}5.04 =$ ________ **24.** $^{+}2\frac{1}{3} + \frac{-1}{3} =$ ________

25. $^{-}3.5 + {}^{+}2.7 + {}^{-}7 =$ ________ **26.** $^{-}1.1 + {}^{-}5.7 + {}^{-}4 =$ ________

27. $^{-}3.8 + {}^{-}2.6 + {}^{+}7.1 =$ ________ **28.** $^{-}8.5 + {}^{-}4.9 + {}^{+}6.1 =$ ________

29. $^{-}5.5 + {}^{+}8.8 + {}^{-}3.3 =$ ________ **30.** $^{+}1.4 + {}^{-}1.2 + {}^{+}0.08 =$ ________

31. $^{-}13.1 + {}^{+}6.1 + {}^{+}1.2 =$ ________ **32.** $^{+}1.1 + {}^{-}9.1 + {}^{+}5 =$ ________

33. $^{-}8.2 + {}^{+}3.3 + {}^{+}5.6 =$ ________ **34.** $^{-}3.5 + {}^{-}7.9 + {}^{+}11.8 =$ ________

Solve.

35. A chemist is working with a certain metal. She changes the temperature as follows: $^{+}1.1$°C, $^{+}2.8$°C, $^{-}1.9$°C. What is the total net change in temperature? ________

36. A hiker climbed 50 m from camp, then went down 21 m, then climbed up another 11 m. Find the net change in his position. ________

*Use with Lesson 4-3, text pages 104–105.

Subtracting Rational Numbers*

Name ____________________

Date ____________________

To subtract a rational number, *add its opposite.* $^{+}7 - {}^{-}3 = \underline{?}$ $\quad {}^{+}7 + {}^{+}3 = {}^{+}10$

Subtract.

1. $^{+}5.1 - {}^{+}6.6 =$ ________ **2.** $^{-}7.3 - {}^{+}3.7 =$ ________ **3.** $^{-}8.1 - {}^{-}5.1 =$ ________

4. $^{+}0.2 - {}^{-}8.01 =$ ________ **5.** $^{-}1.3 - {}^{+}1.1 =$ ________ **6.** $^{-}1.6 - {}^{-}4.6 =$ ________

7. $^{+}14.5 - {}^{-}7.1 =$ ________ **8.** $^{-}13.4 - {}^{+}6.1 =$ ________ **9.** $^{-}10.2 - {}^{+}10.2 =$ ________

10. $^{-}4.5 - {}^{+}4.5 =$ ________ **11.** $^{-}2.8 - {}^{-}2.8 =$ ________ **12.** $^{+}9.8 - {}^{+}0.2 =$ ________

13. $^{+}5 - {}^{-}2\frac{1}{2} =$ ________ **14.** $^{-}4\frac{2}{5} - {}^{-}5\frac{4}{5} =$ ________ **15.** $^{-}1\frac{1}{6} - {}^{+}3\frac{5}{6} =$ ________

16. $^{+}3\frac{7}{10} - {}^{-}5\frac{1}{10} =$ ________ **17.** $^{+}4\frac{1}{2} - {}^{+}3 =$ ________ **18.** $^{-}10\frac{2}{7} - {}^{-}5\frac{1}{7} =$ ________

Simplify each expression. Use the rules for order of operations.

19. $^{-}3.1 - ({}^{+}0.4 - {}^{+}6.1) =$ ________ **20.** $^{+}7.2 + (0 - {}^{+}8.6) =$ ________

21. $^{-}1.1 + ({}^{-}8.7 - {}^{-}0.3) =$ ________ **22.** $^{-}5.05 - ({}^{-}5.1 - {}^{-}5.1) =$ ________

23. $({}^{+}0.39 - {}^{-}0.7) + {}^{-}4.1 =$ ________ **24.** $({}^{+}6.1 - {}^{-}4.4) - ({}^{-}3.3 - {}^{-}2.2) =$ ________

25. $({}^{-}11.1 - {}^{+}0.04) - ({}^{-}2.1 - {}^{+}3.3) =$ ________ **26.** $({}^{+}7.9 + {}^{-}0.10) + ({}^{+}4.1 - {}^{-}6.4) =$ ________

Solve.

27. Mrs. Moffet's special account has a balance of $43.06. She writes a check for $50. What is her new balance? ________

28. A chemist takes a liquid at $^{+}10.4$°C and cools it until its temperature is $^{-}1.5$°C. Find the change in temperature. ________

*Use with Lesson 4-4, text pages 106–107.

Solving Equations*
(Addition and Subtraction)

Name ______________________

Date ______________________

Evaluate each expression.

1. $x - {}^{-}1$ when $x = {}^{+}3$ ______________

2. $x + {}^{-}4$ when $x = {}^{-}2$ ______________

3. $x + {}^{+}5$ when $x = 0$ ______________

4. $x - {}^{+}3$ when $x = {}^{+}6$ ______________

5. ${}^{-}2 + h$ when $h = {}^{+}3$ ______________

6. $a - a$ when $a = {}^{-}2$ ______________

Solve.

7. $s + {}^{+}4 = {}^{+}9$

$s =$ ________

8. $r + {}^{-}7 = {}^{+}14$

$r =$ ________

9. $t + {}^{+}12 = {}^{-}2$

$t =$ ________

10. $b - {}^{+}8 = {}^{-}16$

$b =$ ________

11. $c - {}^{+}11 = {}^{+}42$

$c =$ ________

12. $m - {}^{-}9 = {}^{+}6$

$m =$ ________

13. $c - {}^{-}10 = {}^{+}18$

$c =$ ________

14. $c - {}^{-}2 = {}^{-}1$

$c =$ ________

15. $b + {}^{-}4 = {}^{+}15$

$b =$ ________

16. $d + {}^{-}2.1 = {}^{-}3.4$

$d =$ ________

17. $x - {}^{+}4.1 = {}^{-}2.8$

$x =$ ________

18. $y + {}^{-}2.5 = {}^{-}2.5$

$y =$ ________

19. $a + {}^{+}5 = {}^{+}8$

$a =$ ________

20. $d - {}^{+}7 = {}^{+}21$

$d =$ ________

21. $g - {}^{-}2 = {}^{-}24$

$g =$ ________

Write a mathematical sentence for each. Then solve it.

22. Five less than a number is 26. ______________________________

23. Seven more than a number is ${}^{-}12$. ______________________________

24. A number decreased by ${}^{-}8$ is ${}^{-}11$. ______________________________

25. A number increased by 15 is 41. ______________________________

*Use with Lesson 4-5, text pages 108–109.

Multiplying Rational Numbers*

Name ______________________

Date ______________________

The product of two rational numbers with *different* signs is negative; with the *same* sign is positive.

Find the product.

1. $^{+}3 \times {}^{+}2 =$ ________ **2.** $^{+}6 \times {}^{+}2 =$ ________ **3.** $^{+}3 \times {}^{+}12 =$ ________

4. $^{-}1 \times {}^{-}3 =$ ________ **5.** $^{-}8 \times {}^{-}2 =$ ________ **6.** $^{-}7 \times {}^{-}5 =$ ________

7. $^{-}4 \times {}^{-}7 =$ ________ **8.** $^{-}8 \times {}^{+}3 =$ ________ **9.** $^{+}3 \times {}^{-}4 =$ ________

10. $^{+}5 \times {}^{+}9 =$ ________ **11.** $^{-}10 \times {}^{+}2 =$ ________ **12.** $^{+}6 \times {}^{-}6 =$ ________

13. $^{-}7 \times {}^{-}9 =$ ________ **14.** $^{-}8 \times {}^{-}4 =$ ________ **15.** $^{+}3 \times {}^{-}9 =$ ________

16. $^{-}3.2 \times {}^{-}0.4 =$ ________ **17.** $^{-}0.08 \times {}^{-}0.3 =$ ________ **18.** $^{+}5.1 \times {}^{-}7 =$ ________

19. $^{+}5.14 \times {}^{-}2 =$ ________ **20.** $^{-}4.3 \times {}^{-}0.09 =$ ________ **21.** $^{+}6.07 \times {}^{-}0.01 =$ ________

22. $^{+}5.78 \times {}^{+}2.6 =$ ________ **23.** $-\frac{3}{4} \times -\frac{1}{2} =$ ________ **24.** $+\frac{5}{6} \times -\frac{1}{3} =$ ________

25. $-\frac{1}{4} \times +\frac{3}{4} =$ ________ **26.** $-\frac{3}{4} \times -\frac{8}{9} =$ ________ **27.** $+\frac{2}{9} \times -\frac{5}{6} =$ ________

28. $+\frac{4}{5} \times -\frac{1}{8} =$ ________ **29.** $^{+}2\frac{1}{2} \times -\frac{3}{4} =$ ________ **30.** $^{-}1\frac{1}{8} \times +\frac{1}{2} =$ ________

Compute.

31. $^{-}4 \times ({}^{-}2 + {}^{+}7) =$ ________ **32.** $^{-}7 \times ({}^{-}3 - {}^{-}6) =$ ________

33. $({}^{-}9 - {}^{-}4) \times ({}^{+}3 - {}^{-}1) =$ ________ **34.** $({}^{+}6 + {}^{-}7) \times ({}^{-}4 + {}^{+}4) =$ ________

35. $^{-}5 \times ({}^{-}3 - {}^{-}10) =$ ________ **36.** $({}^{-}8 \times {}^{-}2) + ({}^{-}4 \times {}^{+}2) =$ ________

Evaluate.

37. $6y$ when $y = {}^{-}3$ ________ **38.** $^{-}4t$ when $t = {}^{-}1$ ________

39. $^{-}50m$ when $m = {}^{+}4$ ________ **40.** $^{-}45s$ when $s = 0$ ________

*Use with Lesson 4-6, text pages 110–111.

Dividing Rational Numbers*

Name ______________________

Date ______________________

The quotient of two rational numbers with the *same sign* is positive; with *different signs* is negative.

Complete.

1. $^{+}5 \times$ ________ $= {}^{+}20$

$^{+}20 \div$ ________ $= {}^{+}5$

2. $^{-}8 \times$ ________ $= {}^{+}56$

$^{+}56 \div$ ________ $= {}^{-}8$

3. $^{-}9 \times {}^{-}7 =$ ________

________ $\div {}^{-}7 = {}^{-}9$

4. $^{+}48 \times$ ________ $= {}^{-}96$

$^{-}96 \div$ ________ $= {}^{+}48$

5. $^{-}18 \times$ ________ $= {}^{-}54$

$^{-}54 \div$ ________ $= {}^{-}18$

6. $^{+}12 \times {}^{-}9 =$ ________

________ $\div {}^{-}9 = {}^{+}12$

Divide.

7. $\frac{^{+}72}{^{-}9} =$ ________

8. $\frac{^{-}40}{^{+}8} =$ ________

9. $\frac{^{+}16}{^{-}4} =$ ________

10. $\frac{^{-}21}{^{-}7} =$ ________

11. $\frac{^{-}108}{^{-}12} =$ ________

12. $\frac{^{+}10}{^{+}5} =$ ________

13. $\frac{^{-}56}{^{-}7} =$ ________

14. $\frac{^{+}63}{^{-}7} =$ ________

15. $\frac{^{-}81}{^{+}9} =$ ________

Find the quotient.

16. $^{-}12 \div {}^{+}2 =$ ________

17. $^{-}18 \div {}^{-}9 =$ ________

18. $^{-}25 \div {}^{-}5 =$ ________

19. $^{+}2.3 \div {}^{-}2.3 =$ ________

20. $^{-}3.5 \div {}^{+}0.7 =$ ________

21. $^{+}2.8 \div {}^{-}4 =$ ________

22. $^{-}2\frac{3}{5} \div \frac{^{+}1}{10} =$ ________

23. $^{-}7\frac{1}{4} \div \frac{^{-}5}{8} =$ ________

24. $^{+}5\frac{3}{5} \div {}^{+}7\frac{7}{15} =$ ________

25. $^{-}8\frac{3}{4} \div {}^{+}2\frac{1}{2} =$ ________

26. $^{-}1\frac{1}{8} \div {}^{-}2\frac{3}{4} =$ ________

27. $^{-}6\frac{1}{3} \div {}^{-}4\frac{2}{9} =$ ________

28. $({}^{-}4 \times {}^{-}3) \div {}^{-}2 =$ ________

29. $({}^{-}18 \times {}^{+}3) \div {}^{+}6 =$ ________

30. $({}^{-}28 \times {}^{+}20) \div {}^{-}8 =$ ________

*Use with Lesson 4-7, text pages 112–113.

Solving Equations*
(Multiplication and Division)

Name ____________________

Date ____________________

Write an equation for each. Then solve it.

1. Four times a number is negative 55. ____________________
2. A number divided by negative 3 is positive 7. ____________________
3. A number times negative 9 is negative 36. ____________________
4. Seven times a number is negative 56. ____________________

Solve.

5. $7x = {}^{+}49$ $x =$ ______
6. ${}^{-}5a = {}^{-}45$ $a =$ ______
7. $9r = {}^{+}81$ $r =$ ______
8. ${}^{-}3s = {}^{-}36$ $s =$ ______
9. ${}^{-}8t = {}^{+}0.64$ $t =$ ______
10. ${}^{+}1.4x = {}^{-}42$ $x =$ ______
11. $10y = {}^{-}350.1$ $y =$ ______
12. ${}^{-}0.12a = {}^{+}144$ $a =$ ______
13. $21c = {}^{-}\frac{1}{3}$ $c =$ ______
14. $8d = \frac{4}{5}$ $d =$ ______
15. ${}^{-}23b = {}^{+}20.7$ $b =$ ______
16. $9r = {}^{-}7.65$ $r =$ ______
17. $\frac{1}{4}t = {}^{-}400$ $t =$ ______
18. $13n = {}^{+}91$ $n =$ ______
19. $8x = {}^{-}6$ $x =$ ______
20. $\frac{f}{{}^{+}4} = {}^{+}16$ $f =$ ______
21. $\frac{a}{{}^{+}9} = {}^{-}8$ $a =$ ______
22. $\frac{b}{{}^{-}3} = {}^{+}41$ $b =$ ______
23. $\frac{c}{{}^{+}8} = {}^{-}36$ $c =$ ______
24. $\frac{d}{{}^{+}20} = {}^{-}\frac{1}{5}$ $d =$ ______
25. $\frac{e}{{}^{-}1} = {}^{+}12.5$ $e =$ ______
26. $\frac{h}{{}^{-}1} = {}^{-}4.7$ $h =$ ______
27. $\frac{a}{{}^{-}0.7} = {}^{-}8$ $a =$ ______
28. $\frac{n}{{}^{+}10} = {}^{-}36\frac{1}{5}$ $n =$ ______
29. $\frac{t}{0.5} = {}^{+}26$ $t =$ ______
30. $\frac{c}{{}^{-}4} = {}^{-}10\frac{1}{2}$ $c =$ ______
31. $\frac{n}{{}^{+}3} = {}^{+}64$ $n =$ ______
32. $\frac{x}{{}^{-}2} = {}^{+}5.8$ $x =$ ______

*Use with Lesson 4-8, text pages 114–115.

Equations: Two-Step Solutions*

Name ______________________

Date ______________________

Find the LCM for each equation, when needed. Then solve.

1. $^{-}3.5 + 2x = {}^{-}13.5$ ______

2. $2k - 6 = 8$ ______

3. $24.1 + 6x = 38.5$ ______

4. $\frac{n}{2} - 5 = 15$ ______

5. $\frac{y}{10} + 4 = 16$ ______

6. $\frac{y}{3.5} + 15 = 17$ ______

7. $\frac{s}{3} - 15 = 15$ ______

8. $0.5n + 3 = 8.5$ ______

9. $3y - 11 = 25$ ______

10. $4y - {}^{-}3 = {}^{-}25$ ______

11. $3.6y + 3.2 = 15.8$ ______

12. $\frac{^{-}s}{3} - 4 = {}^{-}10$ ______

Combine like terms. Then solve.

13. $3c + 1\frac{1}{4} - 5c = {}^{-}5$ ______

14. $3a + 2.1 + 4a = {}^{-}0.7$ ______

15. $5r + 2r - 1 = {}^{-}36$ ______

16. $2.1x - 6 + 1.6x = 12.5$ ______

17. $5y - 1 - 3y = {}^{+}5\frac{1}{5}$ ______

18. $9d - 2.3 - 10d = 6$ ______

19. $7x + 1.5 - 8x = {}^{-}2.1$ ______

20. $10c - 20c = 90$ ______

21. $40t - 19t = 231$ ______

Solve equations, using grouping symbols if needed.

22. $\frac{2x - 5 + x - 5}{10} = 8.3$ ______

23. $\frac{n + 4}{4} = 3$ ______

24. $\frac{7}{2.1} = \frac{3}{a}$ ______

25. $5x - 3x - 3 = 4.2$ ______

*Use with Lesson 4-9, text pages 116–117.

Nonroutine Problems*

Name ______________________

Date ______________________

Solve.

1. Each of three cards contains one of the letters A, B, or C, and one of the numbers 1, 2, or 3. A is on the card to the right of C; 3 is on an end card. C and 3 are not on the same card. 2 is on the card to the left of A; 1 is on the card between B and 2. What number and letter are on each card?

2. Carla, Brian, Sula, and Darryl were each born under a different sign of the zodiac. The signs are: Aquarius, Pisces, Leo, and Scorpio. Sula is neither a Scorpio nor a Pisces. Carla is not a Pisces. The name of the Aquarian does not start with the letter B. One of the four is a teacher who is a Leo. Sula, Carla, and Brian are students. What is each person's zodiac sign?

3. Maria, Megan, Ramon, and Jeff are to share in the Pot-O-Gold grand prize. Maria receives $\frac{1}{4}$ of the prize, Ramon $\frac{1}{3}$ of the prize, Jeff $\frac{1}{4}$ of the prize, and Megan receives $72,000. Find each person's share.

4. The distance from town A to town B is 56 mi. Town C is halfway between town A and town B. Town D is $\frac{1}{4}$ of the way between towns A and C, but closer to town C. How far is town A from town D?

5. Moya and Catherine each pick a number from a hat. The difference between 30 and Moya's number is 15. The difference between Catherine's number and 30 is $^{-}15$. Find each number.

*Use for Chapters 1 through 4.

Nonroutine Problems*
(cont'd)

Name ______________________________

Date ______________________________

6. One carton of baseballs contains 10 more baseballs than a second carton. Together they contain 5 dozen baseballs. How many baseballs are in each carton?

7. Juan bought a share of stock. The first week it doubled in value. The second week it decreased $3.50 in value. The third week it increased $5 in value. Juan then sold the share and received $35.50 for it. What was the original cost?

8. Alicia wants to display five albums, equally spaced, on a 5-ft-wide rack. Each album is 9 in. wide. If she leaves no space at the ends, how much space should she allow between each album?

9. Three elephants are used to move some heavy logs blocking a road. Their owners are paid according to the number of tons of logs each elephant moves. The total cost of moving the logs is $693.80. If Mighty Mo moves 5.8 T, Tiny Tim moves 7.3 T, and Sweet Pea moves 6.9 T, how much should each elephant (and owner) receive in payment?

10. A number is divided by 7 and 11 is taken from the quotient. If the difference is 2, what is the number?

*Use for Chapters 1 through 4.

Powers of Ten*

Name ______________________

Date ______________________

Write each power of ten as a standard numeral.

1. 10^3 ________ 2. 10^0 ________ 3. 10^4 ________

4. 10^8 ________ 5. 10^{10} ________ 6. 10^7 ________

Write each as a power of ten (exponent form).

7. $10 \times 10 \times 10 \times 10 \times 10$ ________ 8. 1,000,000,000,000 ________

9. 1 ________ 10. 100 ________

11. 10,000,000 ________ 12. one hundred billion ________

Find the product or quotient.

13. $8.4 \times 10 =$ ________ 14. $0.72 \times 100 =$ ________

15. $9.48 \times 100 =$ ________ 16. $0.347 \times 1000 =$ ________

17. $93 \times 10^2 =$ ________ 18. $8.04 \times 10^3 =$ ________

19. $0.3841 \times 10^3 =$ ________ 20. $8.0315 \times 10^4 =$ ________

21. $6431.1 \times 10^2 =$ ________ 22. $0.3416 \times 10^5 =$ ________

23. $7.93 \div 100 =$ ________ 24. $3.621 \div 1000 =$ ________

25. $50.63 \div 10^2 =$ ________ 26. $0.3141 \div 10^1 =$ ________

27. $1.06 \div 10^5 =$ ________ 28. $8032 \div 10^2 =$ ________

29. $4.2 \div 10^6 =$ ________ 30. $215 \div 10^4 =$ ________

31. $70.09 \div 10^4 =$ ________ 32. $5000 \div 10^7 =$ ________

Arrange in order from greatest to least.

33. 5081.426×10^3; 42.21021×10^5; 6321.004×10^3; 4.321×10^6; 8.22×10^7

____

34. $82.41 \div 10^2$; $3.068 \div 10^1$; $2800 \div 10^4$; $438.61 \div 10^3$; $0.9341 \div 10^0$

____

35. $8.4 \div 10^3$; $8400 \div 10^5$; $0.84 \div 10^6$; $84 \div 10^2$; $8400 \div 10^7$

____

*Use with Lesson 5-1, text pages 128–129.

Negative Exponents*

Name

Date

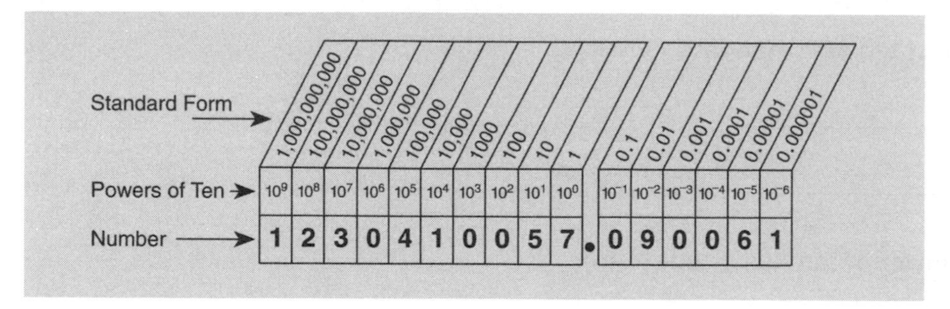

Write the standard numeral.

1. $(8 \times 10^2) + (4 \times 10^1) + (3 \times 10^0) + (7 \times \frac{1}{10^2}) + (4 \times \frac{1}{10^3})$
2. $(3 \times 10^1) + (6 \times 10^{-1}) + (6 \times 10^{-2}) + (4 \times 10^{-3}) + (6 \times 10^{-4})$
3. $(1 \times 10^3) + (2 \times 10^2) + (4 \times 10^0) + (5 \times 10^{-1}) + (6 \times 10^{-2})$
4. $(2 \times 10^2) + (5 \times 10^0) + (3 \times 10^{-2}) + (6 \times 10^{-4}) + (1 \times 10^{-6})$

Give the value of the underlined digit.

5. $3481.2\underline{0}63$
6. $482.\underline{5}1006$
7. $860.514\underline{2}7$
8. $3416.0\underline{0}4$
9. $72{,}106.2007\underline{1}$
10. $928.04\underline{1}65$
11. $3261.9\underline{8}3$
12. $0.426\underline{8}$

Write in expanded form using positive and negative exponents.

13. 0.09
14. 0.004
15. 0.0001
16. 0.000008
17. 66.3
18. 276.8041
19. 838.00052
20. 6000.606
21. 4.000846
22. 203.01050907

*Use with Lesson 5-2, text pages 130–131.

Scientific Notation*
(Multiplication and Division)

Name ______________________

Date ______________________

Complete with the correct power of ten.

1. $782 = 7.82 \times 10^?$ ______

2. $8469 = 8.469 \times 10^?$ ______

3. $0.000052 = 5.2 \times 10^?$ ______

4. $0.00073 = 7.3 \times 10^?$ ______

Write in scientific notation.

5. 62,000 ______

6. 189,000 ______

7. 0.000006 ______

8. 0.000065 ______

Compute. Express the answer in standard notation.

9. $(3.5 \times 10^5) \times (7.2 \times 10^4)$ ______

10. $(5.4 \times 10^3) \times (4.6 \times 10^6)$ ______

11. $(8.1 \times 10^8) \times (2.5 \times 10^2)$ ______

12. $(8.2 \times 10^4) \times (5.5 \times 10^8)$ ______

13. $(9.5 \times 10^2) \times (6.3 \times 10^{-4})$ ______

14. $(1.7 \times 10^{-3}) \times (4.2 \times 10^{-2})$ ______

Divide.

15. $\frac{10^4}{10^1}$ ______

16. $\frac{10^{20}}{10^8}$ ______

17. $\frac{10^0}{10^{-3}}$ ______

18. $\frac{10^6}{10^{-3}}$ ______

19. $6^2 \div 6^{-2}$ ______

20. $5^4 \div 5^8$ ______

21. $8^{-2} \div 8^{-3}$ ______

22. $5^6 \div 5^{-3}$ ______

Compute.

23. $\frac{1.2 \times 10^4}{6 \times 10^2}$ ______

24. $\frac{3.05 \times 10^{10}}{2.5 \times 10^2}$ ______

25. $\frac{6.2 \times 10^{-6}}{5 \times 10^{-3}}$ ______

Solve.

26. The distance of Pluto from the Sun is about 5 910 000 000 km. Write this number in scientific notation. ______

27. The distance of Venus from the Sun is about 107 000 000 km. Write this number in scientific notation. ______

28. If a planet travels 8.9×10^4 mph in its orbit, how far will it travel in 8 days? ______

29. An outer bank rectangular island measures 65 000 m by 9 800 m. What is its area? ______

30. A bird sanctuary in full season has a population of 9,500,000 and an area of 22.5 square miles. What is its population density? ______

*Use with Lessons 5-3, 5-4, text pages 132–135.

Divisibility*

Name ______________________

Date ______________________

Circle the numbers in each row that are divisible by the first number.

1.	3	168	2514	826	43,025	217,251
2.	5	128	450	2645	806,700	324,106
3.	2	97	948	66	27,684	18,630
4.	10	2460	38,155	21,966	34,100	2,671,050
5.	9	324	81,648	32,166	821,940	386,451
6.	8	29,648	321,032	92,611	43,824	592,560
7.	4	247,116	920,424	327,164	807,199	367,148
8.	6	19,800	606,592	408,906	756,101	125,910
9.	11	1,210,117	112,926	617,122	320,693	122,221

Complete each chart. Write "Yes" or "No."

	Number	Divisible by		
		2	5	10
10.	820			
11.	16,325			
12.	73,240			
13.	15,005			
14.	144,240			
15.	255,805			

	Number	Divisible by		
		3	6	9
16.	1269			
17.	5085			
18.	1116			
19.	387,189			
20.	905,436			
21.	539,514			

22. Write the smallest number divisible by 2, 5, and 10. ______________________

23. Write the smallest number divisible by 3, 6, and 9. ______________________

24. Write the smallest number divisible by 2, 3, 4, 6, and 8. ______________________

25. Mr. Martin baked 688 fresh doughnuts. He wants to package them in boxes of 8. Will each box be filled with doughnuts? How many boxes will he use? ______________________

*Use with Lesson 5-5, text pages 136–137.

Patterns and Sequences*

Name ______________________

Date ______________________

Find the next three terms in each sequence.

1. 4, 9, 14, 19, ______________

2. 1, 4, 7, 10, ______________

3. 54, 52, 50, 48, ______________

4. 2, 11, 20, 29, ______________

5. 1, 2, 4, 8, ______________

6. 1600, 1500, 1400, 1300, ______________

7. 8, 15, 22, 29, ______________

8. 7, 13, 19, 25, ______________

9. 10, 10.5, 11, 11.5 ______________

10. 27, 38, 49, 60, ______________

11. $\frac{1}{3}$, $\frac{2}{3}$, 1, $\frac{4}{3}$, ______________

12. 96, 93, 90, 87, ______________

13. 1, 2, 4, 7, ______________

14. 100, 95, 94, 89, ______________

15. 12.3, 12.5, 12.7, 12.9, ______________

16. 1, 3, 2, 4, ______________

17. 5, 6, 11, 12, ______________

18. 2, 7, 5, 10, 8, ______________

19. 90, 80, 71, 63, ______________

20. 1, $\frac{1}{2}$, $\frac{1}{4}$, $\frac{1}{8}$, ______________

21. 7, 10, 13, 16, ______________

22. 92, 88, 84, 80, ______________

23. 1, 3, 9, 27, ______________

24. 2048, 1024, 512, 256, ______________

25. 1, 1.3, 1.6, 1.9, ______________

26. 7, 20, 33, 46, ______________

27. 92, 91, 89, 86, ______________

28. 2, 6, 5, 9, ______________

29. 4, 5.5, 7, 8.5, ______________

30. 13, 16, 11, 14, ______________

31. 30, 15, 18, 9, ______________

32. 70, 210, 110, 330, ______________

33. 8, 32, 16, 64, ______________

34. 10,000, 1000, 100, 10, ______________

***Use with Lesson 5-6, text page 138.**

Connectives*

Name ______________________

Date ______________________

Use *p*, *q*, and *r* to write each connective in symbols where:

***p*:** I seal the letter. ***q*:** You open the book. ***r*:** Nicki watches TV.

1. I seal the letter and you open the book. ______________________
2. You open the book or I seal the letter. ______________________
3. If you open the book, Nicki watches TV. ______________________
4. I do *not* seal the letter and Nicki does *not* watch TV. ______________________
5. If Nicki watches TV, then I seal the letter and you open the book. ______________________
6. It is not true that I seal the letter and you open the book. ______________________

Use *r*, *s*, and *t* to write a statement in these exercises where:

***r*:** Love is everything. ***s*:** Silence gives consent. ***t*:** Time is a thief.

7. $r \wedge s$ ______________________
8. $r \vee \sim t$ ______________________
9. $s \rightarrow t$ ______________________
10. $s \leftrightarrow t$ ______________________
11. $t \vee r$ ______________________
12. $r \rightarrow (s \wedge t)$ ______________________

Complete these truth tables.

13.

p	*q*	$\sim p$	$\sim p \leftrightarrow q$
T	T		
T	F		
F	T		
F	F		

14.

p	*q*	$\sim q$	$p \wedge \sim q$
T	T		
T	F		
F	T		
F	F		

15.

p	*q*	$\sim p$	$\sim p \wedge q$
T	T		
T	F		
F	T		
F	F		

Construct a truth table for each.

16. $\sim p \vee \sim q$ ______________
17. $\sim(p \vee q)$ ______________
18. $(p \vee q) \rightarrow (p \wedge q)$ ______________

*Use with Lessons 5-7, 5-8, text pages 139–141.

Converse, Inverse, Contrapositive*

Name ______________________

Date ______________________

Write the converse, inverse, and contrapositive for each conditional.

1. If a number is not divisible by one, it is zero.

____

____

____

2. If Killer is a dog, then Killer is a beagle.

____

____

____

3. If Stretch is a python, then Stretch is a snake.

____

____

____

Determine the truth value of each conditional, its converse, its inverse, and its contrapositive. The first one is done for you.

4. If water freezes at 0° C, then $2 + 3 = 5$. T, T, T, T

5. If the World Trade Center has 104 floors, then $2 \div 2 = 1$. __________

6. If 1996 is a leap year, then February will have 29 days. __________

7. If Boston is in Massachusetts, then the ice is hot. __________

8. If sixteen is divisible by 5, then 7 is an even number. __________

9. If $3 + 3 \neq 6$, then 109 is divisible by 3. __________

Complete this truth table.

10.

p	q	$q \rightarrow p$	$\sim q$	$\sim p$	$(\sim p \rightarrow \sim q)$	$(q \rightarrow p) \rightarrow (\sim p \rightarrow \sim q)$
T	T					
T	F					
F	T					
F	F					

*Use with Lesson 5-9, text pages 142–143.

Ratios and Rates*

Name ______________________

Date ______________________

Write a ratio. Express it as a fraction in lowest terms. Change to like units where necessary.

1. 9 out of 10 people use Sure Glue ________

2. 25¢ out of every dollar goes to entertainment ________

3. 4 girls in a class of 9 students ________

4. The number of even digits to the number of odd digits in 9,241,657 ________

5. 8 apples in a basket of 2 dozen ________

6. 6 inches out of a yard ________

Write as a rate in simplest form.

7. 3 pages in 5 hours ________

8. 7 books in 8 weeks ________

9. $450 a month ________

10. 6 cans for $1.29 ________

11. 2 shirts for $12.50 ________

12. 12 guppies for $6.39 ________

13. 500 yards in 40 seconds ________

14. 6 bags for $3.50 ________

Write an equal ratio in lowest terms.

15. 30 : 5 ________

16. 7 : 49 ________

17. 6 : 24 ________

18. 40 : 8 ________

19. 45 : 81 ________

20. 24 : 72 ________

21. 50 : 25 ________

22. 36 : 54 ________

23. 24 : 48 ________

24. 42:12 ________

25. $4\frac{1}{5} : 3\frac{1}{2}$ ________

26. $\frac{2}{3} : \frac{6}{9}$ ________

27. $1\frac{1}{3} : 2\frac{5}{6}$ ________

28. $\frac{2}{3} : \frac{3}{4}$ ________

29. $3\frac{1}{5} : 2\frac{2}{3}$ ________

30. $3\frac{1}{3} : \frac{10}{15}$ ________

31. 1.6 : 0.4 ________

32. 8.1 : 1.8 ________

33. 6.4 : 1.6 ________

34. 2.5 : 6.25 ________

35. 7.2 : 0.8 ________

Solve.

36. Justin is 7 years old and his granddad is 77. What is the ratio of Justin's age to his granddad's age? ________

37. What is the ratio of the number of vowels to the number of consonants in this question? ________

*Use with Lesson 6-1, text pages 154–155.

Proportions*

Name ______________________

Date ______________________

Write = or ≠.

1. $\frac{7}{9}$ ____ $\frac{21}{27}$
2. $\frac{28}{35}$ ____ $\frac{4}{5}$
3. $\frac{8}{7}$ ____ $\frac{96}{49}$
4. $\frac{46}{58}$ ____ $\frac{23}{29}$
5. 6 : 138 ____ 4 : 92
6. 12 : 60 ____ 4 : 24
7. 10 : 90 ____ 45 : 5
8. 9 : 27 ____ 3 : 9
9. $\frac{4.2}{5.6}$ ____ $\frac{0.6}{0.8}$
10. $\frac{1.44}{1.32}$ ____ $\frac{1.2}{1.1}$
11. $\frac{8\frac{1}{2}}{5}$ ____ $\frac{25\frac{1}{2}}{7\frac{1}{2}}$
12. $\frac{2\frac{1}{3}}{1\frac{2}{5}}$ ____ $\frac{11\frac{2}{3}}{7}$

Solve.

13. $\frac{27}{63} = \frac{n}{7}$ n = ________
14. $\frac{48}{x} = \frac{6}{12}$ x = ________
15. $\frac{a}{15} = \frac{18}{45}$ a = ________
16. $\frac{21}{b} = \frac{147}{105}$ b = ________
17. $\frac{42}{24} = \frac{a}{4}$ a = ________
18. $\frac{72}{64} = \frac{18}{r}$ r = ________
19. $\frac{\frac{5}{9}}{1\frac{1}{4}} = \frac{\frac{2}{3}}{s}$ s = ________
20. $\frac{b}{\frac{1}{3}} = \frac{4}{2}$ b = ________
21. $\frac{4.8}{1.32} = \frac{c}{1.1}$ c = ________
22. $\frac{9.6}{a} = \frac{0.8}{0.6}$ a = ________
23. $\frac{n}{0.24} = \frac{0.05}{0.03}$ n = ________
24. $\frac{2.4}{3.7} = \frac{1.2}{t}$ t = ________
25. $\frac{r}{2\frac{1}{2}} = \frac{1}{2}$ r = ________
26. $\frac{2}{b} = \frac{\frac{5}{6}}{\frac{1}{3}}$ b = ________
27. $\frac{s}{2\frac{1}{2}} = \frac{2\frac{1}{2}}{1\frac{1}{4}}$ s = ________
28. $\frac{\frac{3}{7}}{\frac{2}{3}} = \frac{\frac{1}{4}}{c}$ c = ________
29. $\frac{a}{0.25} = \frac{0.8}{0.5}$ a = ________
30. $\frac{3.6}{c} = \frac{1.2}{2.1}$ c = ________

31. If 2.5 cm on a map represents 10 kilometers, how many kilometers will be represented by 4.8 cm? ______________________

*Use with Lesson 6-2, text pages 156–157.

Direct Proportion*

Name ____________________

Date ____________________

Solve by proportion. Round to the nearest cent where necessary.

1. An ad says that 6 lemons cost 94¢. What will 8 lemons cost? ____________________

2. Strawberries are marked 20 oz for $1.29. How much will 15 oz cost? ____________________

3. A package of paper plates holding 150 plates costs $3.29. At that rate, how many would be in a package that cost $4.72? ____________________

4. Hot dog rolls are priced at 2 packages for $1.12. How many packages can be bought for $3.86? ____________________

5. A 30-oz can of lemonade drink mix costs $1.80. At that rate, how much will a 21-oz can cost? ____________________

6. If a 30-oz can of lemonade drink mix makes 8 quarts, how much mix is needed to make 10 quarts of the drink? ____________________

7. Olives are 69¢ for $5\frac{3}{4}$ ounces. At this rate, what will $2\frac{1}{2}$ ounces of olives cost? ____________________

8. A six pack of the store-brand cola cost $1.64. How much must Denise pay for 14 cans of it? ____________________

9. If a car travels 360 km in 5.4 hours, how far will it travel in 7.8 hours? ____________________

10. If 12 tickets for a football game cost $98.40, how many can be bought for $73.80? ____________________

11. If an office worker earns $9460 in $\frac{4}{5}$ of a year, how much will he earn in $\frac{1}{5}$ of a year? ____________________

12. Soup sells for 3 cans for 99¢. How much will 15 cans cost? ____________________

*Use with Lesson 6-3, text pages 158–159.

Scale Drawings*

Name ______________________

Date ______________________

Complete. Write and solve each proportion using the given scale.

	Scale: 2 cm = 5 km	
	Measures:	
	Scale	Actual
1.	4 cm	
2.		12 km
3.		75 km
4.	1 cm	
5.		40 km
6.	0.5 cm	
7.		110 km

	Scale: 0.3 cm = 1.7 km	
	Measures:	
	Scale	Actual
8.	2.4 cm	
9.		11.9 km
10.	2.7 cm	
11.		0.425 km
12.	0.06 cm	
13.	2.16 cm	
14.		17 km

15. **Find the scale used to make this plan. Then label all the sides with their actual measures. Write the actual measures in the boxes.**

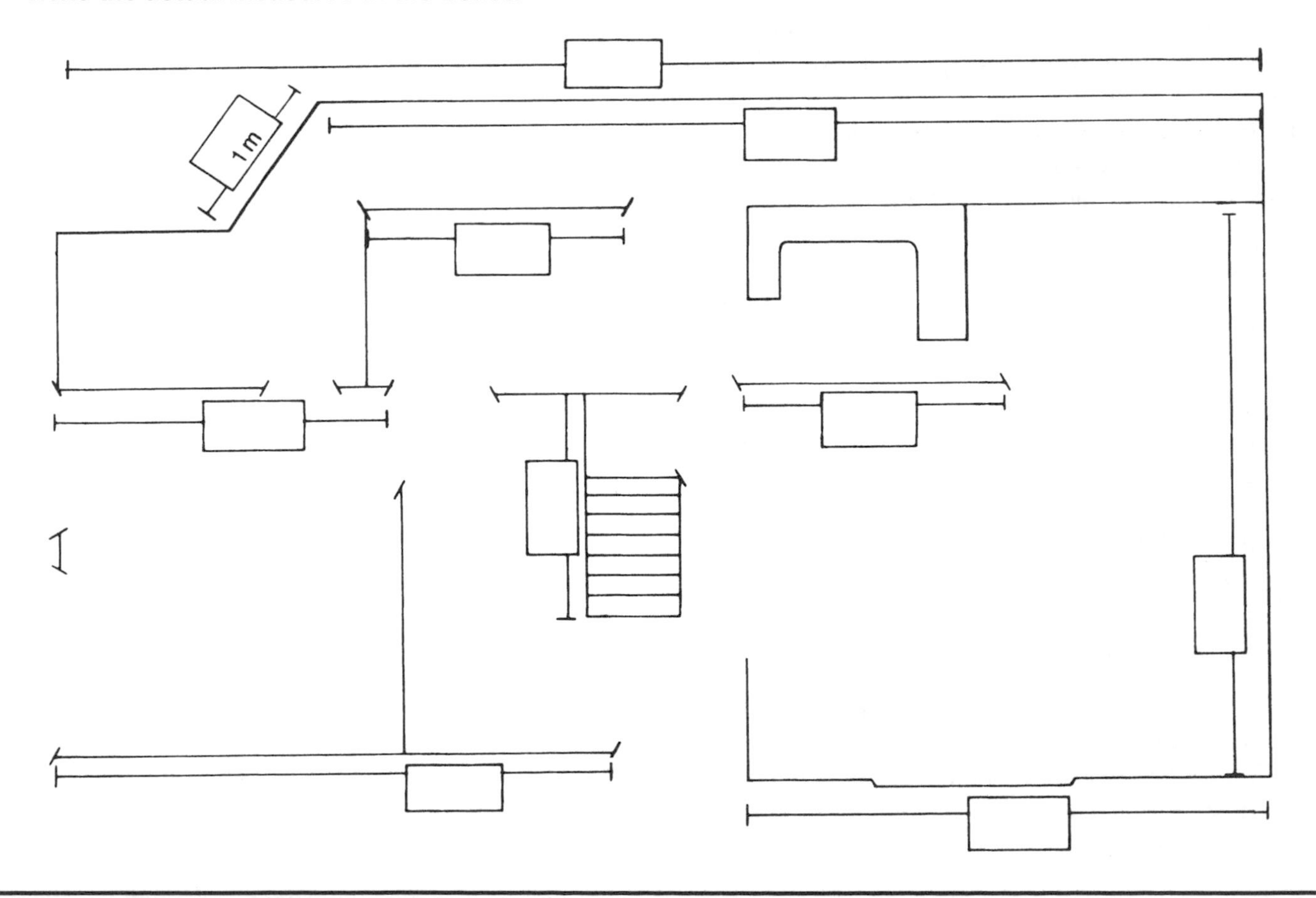

Inverse Proportion*

Name ______________________

Date ______________________

Write a proportion. Solve.

1. 4 bricklayers build a wall in 5 days.

   10 bricklayers build it in _______ days.

2. 6 students decorate the gym in 3 hours.

   2 students do it in _______ hours.

3. 8 workers plant a garden in 4 hours.

   16 workers plant it in _______ hours.

4. 7 campers put up a tent in 45 minutes.

   4 campers put it up in _______ minutes.

5. 5 people deliver the papers in $1\frac{1}{2}$ hours.

   _______ people deliver the papers in $\frac{3}{4}$ hour.

6. 3 students made the scenery in 8 days.

   _______ students can make it in 4 days.

Solve.

7. If 8 people do a piece of work in 12 days, how many people will be needed to do the same work in 16 days? ______________________

8. If 3 workers can build a deck in $5\frac{2}{3}$ days, how many days will it take 5 workers to build the same deck? ______________________

9. It took 7 hours for 12 doctors to give routine physical exams to 204 patients. How long would it have taken if 8 doctors gave the exams? ______________________

10. How many students would be needed to distribute 780 advertisements in 2 hours if 9 students can distribute 540 ads in 2 hours? ______________________

11. Two people decorated 10 cakes in $2\frac{1}{2}$ hours. How many hours would it have taken if 5 people had been used to decorate the cakes? ______________________

12. Three workers can repave a driveway in 5 hours. How long will it take 8 workers to repave the driveway? ______________________

13. It took 2 hours for 10 students to write 240 invitations to the school carnival. How long would it have taken if 5 students had done the writing? ______________________

14. Four postal workers can sort 72 packages in $2\frac{1}{2}$ hours. If 2 extra workers are hired, how much time would be needed to sort the same number of packages? ______________________

*Use with Lesson 6-5, text pages 162–163.

Partitive Proportion*

Name ______________________

Date ______________________

Solve using partitive proportion.

1. Divide 64 into two parts with a ratio of 1 to 3. ______________________
2. Divide 600 into two parts with a ratio of 10 to 20. ______________________
3. Divide 270 into three parts with a ratio of 2 : 3 : 4. ______________________
4. Divide 300 into three parts with a ratio of 4 : 5 : 6. ______________________
5. Divide 450 into three parts with a ratio of 4 : 5 : 6. ______________________
6. Divide 810 into three parts with a ratio of 2 : 3 : 4. ______________________

Solve.

7. Three friends divided 165 baseball cards among themselves in a ratio of 1 : 4 : 6. How many cards did each receive? ______________________
8. A toy manufacturer made 1424 stuffed animals. They made teddy bears, bunnies, and dogs in the ratio of 8 : 5 : 3. How many of each kind of stuffed animal did they make? ______________________
9. John earned $6 for every $4 that Martin earned. How much did each earn if they earned $75 together? ______________________
10. The main library purchased 938 new books. They want to distribute the new books to three of their branch libraries in the ratio of 3 : 5 : 6. How many books will be given to each library? ______________________
11. A card store ordered 520 cards. They ordered birthday cards, anniversary cards, and get well cards in the ratio of 6 : 3 : 1. How many of each type of card did they order? ______________________
12. At a nursery, there are 1248 flowers. There are roses, orchids, and carnations in the ratio of 3 : 4 : 6. How many of each kind of flower are there? ______________________
13. In a local election, 2640 votes were cast for two candidates. Ms. Wayne received 7 votes for every 4 votes that Mr. Edwards received. How many votes did each candidate receive? ______________________
14. A painter bought 96 L of paint. He bought blue, green, and yellow paint in the ratio of 4 : 3 : 1. How many liters of each color paint did he buy? ______________________

*Use with Lesson 6-6, text pages 164–165.

Similar Triangles and Trigonometric Ratios*

Name ______________________

Date ______________________

The figures below are similar. Write and solve proportions to find the indicated sides.

1.

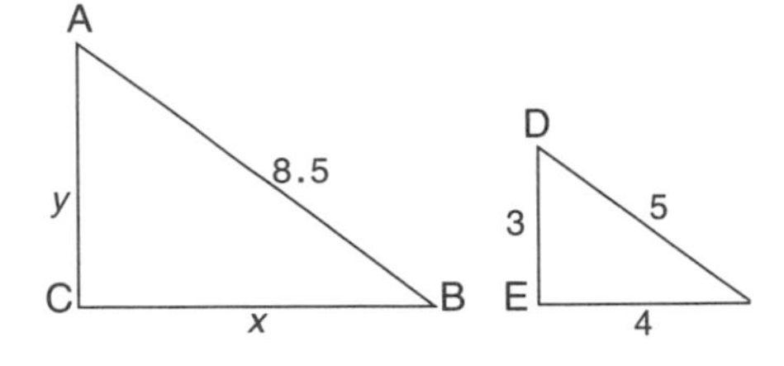

2.

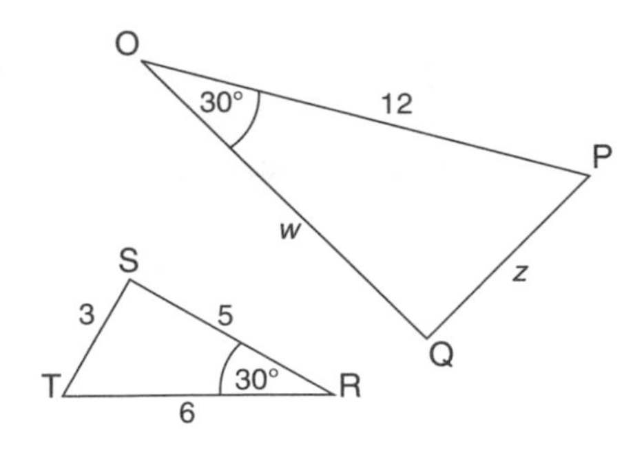

Use similar triangles *EFG* and *HJK* to solve.

3. $\frac{\overline{FG}}{\overline{JK}} = \frac{\overline{EG}}{\_\_\_\_}$

4. $\frac{\overline{HJ}}{\overline{EF}} = \frac{\overline{KH}}{\_\_\_\_}$

5. $\angle F \cong \angle J =$ ______

6. $\frac{x}{11} = \frac{7.5}{10}$

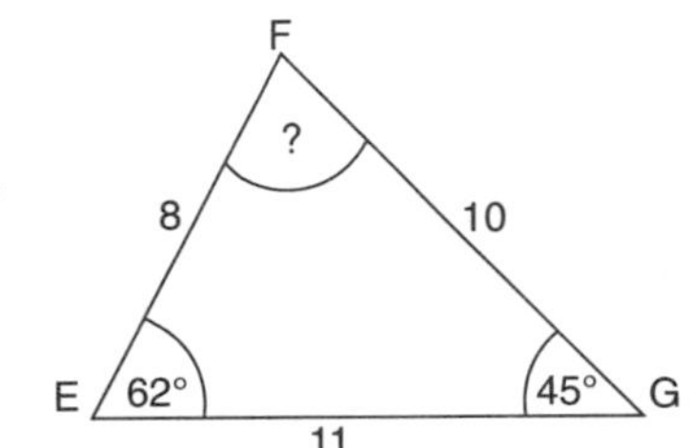

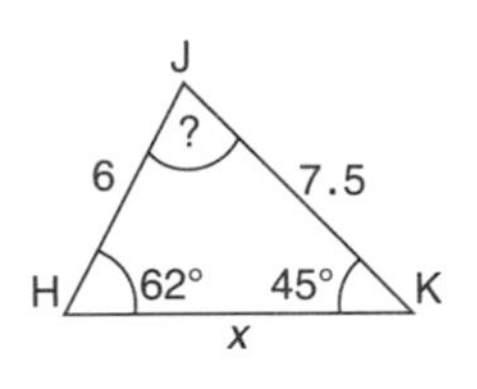

Find the value of each trigonometric ratio. Then use the table of trigonometric ratios (page 196) to find the measure of the angle.

(page 196)

7. tan B **8.** sin A **9.** cos B

10. sin B **11.** cos A **12.** tan A

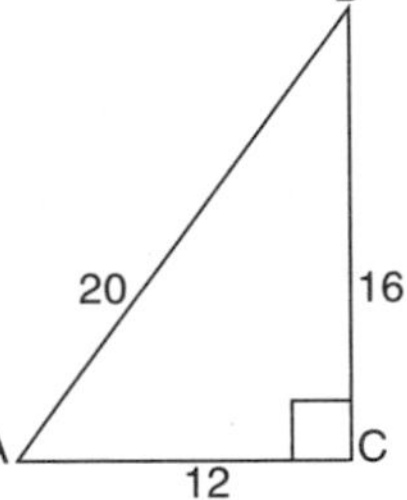

Solve, using a scientific calculator.

13. Find the length of side s.

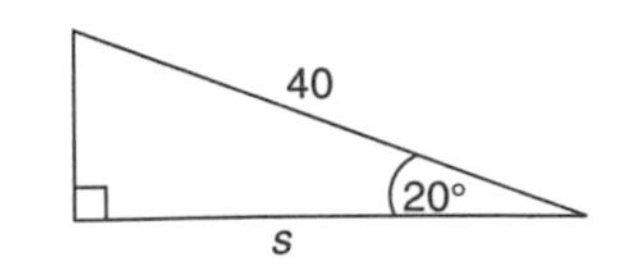

14. Find the length of side r.

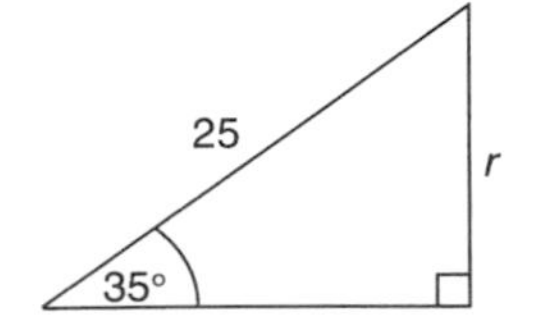

15. Find the length of side h.

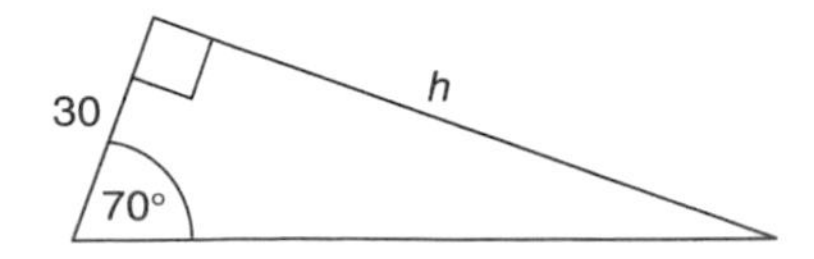

16. Find the length of side t.

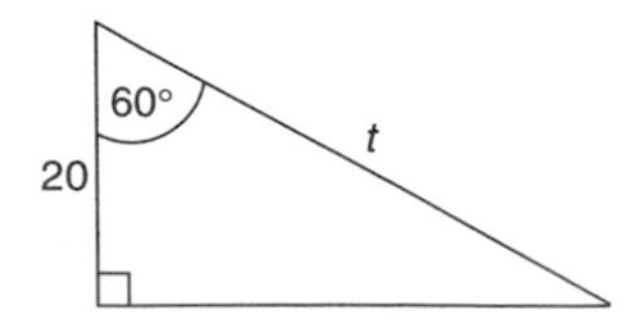

*Use with Lessons 6-7, 6-8, text pages 166–169.

Fractions and Decimals to Percents*

Name ____________________

Date ____________________

Complete the chart.

	Ratio	Fraction	Decimal	Percent
1.	30:100	$\frac{3}{10}$		
2.	50:100			
3.		$\frac{1}{4}$		
4.	$16\frac{2}{3}$:100	$\frac{1}{6}$		
5.			0.8	
6.		$\frac{1}{3}$		
7.			0.42	
8.			0.9	
9.	5:8			
10.		$\frac{7}{9}$		
11.			0.6	
12.		$1\frac{3}{4}$		
13.		$\frac{1}{8}$		
14.	2:3			

Circle the percent, ratio, fraction, or decimal that does not belong in each exercise.

15. 2.5 25% $\frac{1}{4}$ 25:100

16. 13% 0.13 $\frac{13}{100}$ 1.3:100

17. 10:100 $\frac{70}{100}$ 70% 0.7

18. 1:2 50% $\frac{50}{100}$ 0.05

19. 45% 9:20 $\frac{45}{100}$ 4.5

20. 35% 35:100 0.35 $\frac{7}{10}$

21. 0.375 $37\frac{1}{2}$% $\frac{3}{8}$ 37:100

22. 1:3 $0.66\frac{2}{3}$ $\frac{2}{3}$ $66\frac{2}{3}$%

23. 2% 0.02 2:100 $\frac{2}{50}$

24. 48:100 48% $\frac{12}{50}$ 0.48

25. 0.15 $\frac{150}{100}$ 150% 150:100

26. 99% 99:100 $\frac{9}{10}$ 0.99

27. 0.875 87:100 $87\frac{1}{2}$% $\frac{7}{8}$

28. 0.3 $\frac{3}{10}$ 30:100 3%

*Use with Lesson 7-1, text pages 180–181.

Percents to Fractions and Decimals*

Name ______________________

Date ______________________

Change to a fraction in lowest terms.

1. 40% = ______ **2.** 80% = ______ **3.** 32% = ______ **4.** 48% = ______

5. $12\frac{1}{2}\%$ = ______ **6.** $83\frac{1}{3}\%$ = ______ **7.** $16\frac{2}{3}\%$ = ______ **8.** $22\frac{2}{9}\%$ = ______

9. $14\frac{2}{7}\%$ = ______ **10.** $8\frac{1}{3}\%$ = ______ **11.** 62.5% = ______ **12.** $66\frac{2}{3}\%$ = ______

13. 125% = ______ **14.** 120% = ______ **15.** 275% = ______ **16.** 200% = ______

17. 4.5% = ______ **18.** 2.25% = ______ **19.** 3.5% = ______ **20.** 6.25% = ______

21. 0.4% = ______ **22.** 0.25% = ______ **23.** 105% = ______ **24.** 335% = ______

Change to a decimal.

25. 65% = ______ **26.** 46% = ______ **27.** 98% = ______ **28.** 72% = ______

29. 24% = ______ **30.** 6% = ______ **31.** 4% = ______ **32.** 8% = ______

33. 35.5% = ______ **34.** 86.2% = ______ **35.** 75.1% = ______ **36.** 13.2% = ______

37. 2.25% = ______ **38.** 3.6% = ______ **39.** 7.4% = ______ **40.** 8.6% = ______

41. 100% = ______ **42.** 200% = ______ **43.** 250% = ______ **44.** 325% = ______

45. 0.6% = ______ **46.** 0.25% = ______ **47.** 0.4% = ______ **48.** 0.15% = ______

Solve.

49. The scouts sold 40 of the 4 dozen cakes baked for the fund-raising event. What percent of the cakes did they sell? ______________________

50. 0.85 of the seats for the 3:45 flight to Phoenix were filled. What percent of the seats were not filled? ______________________

*Use with Lesson 7-2, text pages 182–183.

Finding a Part or a Percentage*

Name ______________________

Date ______________________

Find the part or percentage of each number.

1. 10% of 40 = ________

2. 75% of 16 = ________

3. 50% of 42 = ________

4. 30% of 90 = ________

5. 5% of 70 = ________

6. 20% of 100 = ________

7. 80% of 10 = ________

8. 2% of 14 = ________

9. 36% of 78 = ________

10. 24% of 28 = ________

11. 1% of 34 = ________

12. 60% of 240 = ________

13. $12\frac{1}{2}$% of 56 = ________

14. $33\frac{1}{3}$% of 48 = ________

15. $16\frac{2}{3}$% of 246 = ________

16. 25% of 316 = ________

17. $87\frac{1}{2}$% of 32 = ________

18. $66\frac{2}{3}$% of 543 = ________

19. $3\frac{1}{4}$% of 620 = ________

20. 400% of 85 = ________

21. 0.75% of 30 = ________

Solve.

22. In order to determine the need for traffic signals at a suburban intersection, a 48-hour watch was kept. During that time, $12\frac{1}{2}$% of the 8424 cars recorded made left-hand turns. How many cars was this?

____

23. During the same watch, it was noted that $16\frac{2}{3}$% of the cars made right-hand turns. How many cars made right-hand turns?

____

24. Last summer, Julie earned $145. This summer, she earned 25% more. How much did she earn this summer?

____

25. 30% of 50 students play a musical instrument. How many students play musical instruments?

____

*Use with Lesson 7-3, text pages 184–185.

Finding Percent*

Name ______________________

Date ______________________

Find the percent. Round to the nearest tenth, if necessary.

1. What percent of 8 is 2? ____________

2. What percent of 9 is 3? ____________

3. What percent of 25 is 5? ____________

4. 8 is what percent of 40? ____________

5. 12 is what percent of 50? ____________

6. What percent of 16 is 48? ____________

7. What percent of 120 is 80? ____________

8. What percent of 9 is 72? ____________

9. 11 is what percent of 99? ____________

10. What percent of 48 is 36? ____________

11. What percent of 14 is 49? ____________

12. 54 is what percent of 90? ____________

13. What percent of 24 is 2? ____________

14. What percent of 500 is 600? ____________

15. 81 is what percent of 36? ____________

16. What percent of 963 is 642? ____________

17. What percent of 455 is 182? ____________

18. What percent of 224 is 140? ____________

Solve.

19. A team won 18 games and lost 9 games. The number of games won is what percent of the total games played? ____________

20. Bob earned $810 last summer. He spent $540. What percent of his earnings did he save? ____________

***Use with Lesson 7-4, text pages 186–187.**

Finding the Original Number*

Name ______________________

Date ______________________

75% of n = 27   75% × n = 27 n = 27 ÷ 75% n = 36

Find the original number.

1. 3% of a = 42 a = ________

2. 10% of b = 9 b = ________

3. 50% of c = 16 c = ________

4. 20% of d = 10 d = ________

5. 80% of e = 12 e = ________

6. 25% of f = 20 f = ________

7. 75% of g = 18 g = ________

8. 12% of h = 3.6 h = ________

9. $33\frac{1}{3}$% of i = 17 i = ________

10. 45% of j = $20\frac{1}{4}$ j = ________

11. 16% of k = 8.32 k = ________

12. 2% of l = 8.4 l = ________

13. $16\frac{2}{3}$% of m = 44 m = ________

14. $12\frac{1}{2}$% of n = 70 n = ________

15. 2.5% of o = $12\frac{1}{2}$ o = ________

16. 1.8% of p = 0.9 p = ________

17. 150% of q = 7.5 q = ________

18. 300% of r = 12 r = ________

Complete the table.

a = rate; b = percentage; c = original number or base

	a	b	c
19.	14%	0.84	
20.	30%		15
21.		3	21
22.	25%	100	
23.	$66\frac{2}{3}$%		81
24.		9	45
25.	80%		30
26.	$16\frac{2}{3}$%	6	

*Use with Lesson 7-5, text pages 188–189.

Estimating with Percents*

Name ______________________

Date ______________________

Use your Table of Common Percents (p. 562) to estimate each percent.

1. 32% ______ 2. 48% ______ 3. 13% ______ 4. 19% ______ 5. 33% ______

6. $9\frac{2}{3}$% ______ 7. 41.5% ______ 8. 76.1% ______ 9. $18\frac{2}{3}$% ______ 10. 81.9% ______

Estimate the percentage.

11. 9% of 71 ______ 12. 53% of 98.6 ______ 13. 42% of 38.9 ______

14. $6\frac{7}{8}$% of 16 ______ 15. 92% of 30 ______ 16. 14.5% of 76 ______

Choose the best ratio and then estimate each percent.

17. ______ % of 22 is 8. a. $\frac{9}{15}$ b. $\frac{7}{21}$ c. $\frac{22}{8}$

18. ______ % of 89 is 11.6. a. $\frac{11}{88}$ b. $\frac{10}{100}$ c. $\frac{90}{100}$

19. ______ % of 14.4 is 9.2. a. $\frac{8}{15}$ b. $\frac{9}{15}$ c. $\frac{10}{20}$

Estimate the original number.

20. 9% of ______ is 7. 21. 36.5% of ______ is 42.

22. 7.8% of ______ is 18. 23. 14% of ______ is 21.

Solve.

24. Of 526 students taking a high school entrance exam, 6.5% scored in the top percentile. About how many students were in that percentile? ______

25. A survey of 12,212 people in four states showed that 3.6% were not able to read. About how many people is that? ______

*Use with Lesson 7-6, text pages 190–191.

Percent of Increase or Decrease*

Name ______________________

Date ______________________

Complete. (Round each answer to the nearest tenth of a percent.)

Percent of Change in Car Sales

		Car Sales		Increase		Decrease	
	Month	1981	1982	Number	%	Number	%
1.	March	45	50				
2.	April	64	53				
3.	May	60	52				
4.	June	59	62				
5.	July	58	65				
6.	August	60	51				
7.	September	74	80				
8.	October	61	51				
9.	November	56	62				
10.	December	30	34				

Solve. (Round each answer to the nearest tenth of a percent.)

11. Last year, 11 students tried out for the debating team. This year, 15 students tried out. What was the percent of increase? ______________________

12. 11 inches of rain fell last month. This month, 10 inches fell. What was the percent of decrease? ______________________

13. Sales of radios decreased one month from 150 to 100. What was the percent of decrease? ______________________

14. Joan worked 28 hours last week and 34 hours this week. What is the percent of increase in her work time? ______________________

15. In last year's graduating class, 75 students went to college. From this year's class, 80 students will go to college. Find the percent of change. ______________________

16. 1450 new texts have been purchased for this school year. Last year, 1600 were purchased. What was the percent of change? ______________________

*Use with Lesson 7-7, text pages 192–193.

Progress Test: Chapters 1–7*

Name ______________________

Date ______________________

Write each answer on the numbered line. Do all computations on a separate sheet of paper.

Answers

Choose the correct answer. Write the letter.

1. One hundred billion, four hundred million, fifty-five thousand four is:
a. 50,004,004 **b.** 100,400,055,004 **c.** 100,055,004 **d.** 554,000,400 — **1.** ______ (1–1)

2. Rounded to the nearest thousandth, 624.0729 is:
a. 624.1 **b.** 62.507 **c.** 624.073 **d.** 624.072 — **2.** ______ (1–3)

3. The best estimate for 461 + 1739 + 23,104 is:
a. 24,200 **b.** 25,000 **c.** 25,300 **d.** 24,000 — **3.** ______ (1–9)

4. $31.213 \div 4.9$ rounded to the nearest tenth is:
a. 0.6 **b.** 6.3 **c.** 6.4 **d.** 6.37 — **4.** ______ (1–7)

5. If $n = 9$, then the value of $n - 9$ is:
a. 18 **b.** 1 **c.** 0 **d.** 81 — **5.** ______ (3–3)

6. 3 more than 4 times a number (a) is 27.
To solve for a, use the equation:
a. $3a + 4 = 27$ **b.** $3a - 4 = 27$ **c.** $4a + 3 = 27$ **d.** $4a - 3 = 27$ — **6.** ______ (3–2)

7. The decimal equivalent of $\frac{7}{8}$ is:
a. 0.78 **b.** 0.785 **c.** 0.875 **d.** 0.87 — **7.** ______ (2–7)

8. $\frac{7}{12}$ of 15 means:
a. $\frac{7}{12} + 15$ **b.** $\frac{7}{12} - 15$ **c.** $\frac{7}{12} \times 15$ **d.** $\frac{7}{12} \div 15$ — **8.** ______ (2–11)

9. $4\frac{5}{9} - 2\frac{2}{3}$ equals:
a. $2\frac{1}{3}$ **b.** $2\frac{8}{9}$ **c.** $2\frac{1}{9}$ **d.** $1\frac{8}{9}$ — **9.** ______ (2–10)

10. $\frac{3}{5} \times \frac{1}{6} + \frac{2}{5}$ equals:
a. $1\frac{1}{6}$ **b.** $\frac{1}{2}$ **c.** $\frac{1}{3}$ **d.** 4 — **10.** ______ (2–11)

11. The reciprocal of $2\frac{2}{3}$ is:
a. $\frac{8}{3}$ **b.** $2\frac{1}{2}$ **c.** $\frac{3}{8}$ **d.** $\frac{3}{22}$ — **11.** ______ (2–12)

12. If $n - {}^{-}4 = 6$, then n equals:
a. $^{-}2$ **b.** $^{+}2$ **c.** $^{-}10$ **d.** $^{+}10$ — **12.** ______ (3–13)

13. To solve $\frac{3}{4} \times n = 30$:
a. divide both sides by 30 **b.** multiply both sides by 30
c. divide both sides by $\frac{3}{4}$ **d.** multiply both sides by $\frac{3}{4}$ — **13.** ______ (3–7)

14. Choose the composite number equal to $2^3 \times 5 \times 7$.
a. 125 **b.** 96 **c.** 280 **d.** 210 — **14.** ______ (2–4)

15. The fraction equivalent to $14\frac{2}{7}\%$ is:
a. $\frac{4}{7}$ **b.** $\frac{2}{7}$ **c.** $\frac{1}{14}$ **d.** $\frac{1}{7}$ — **15.** ______ (7–2)

***Next to each item is given the lesson number in the text where the item was taught.**

Progress Test: Chapters 1–7*

Name ________________________

Date ________________________

Compare. Write $<$, $=$, or $>$.

16. $0.023 \underline{\ ?\ } 0.23$

17. $12(1.6 - 0.6) \underline{\ ?\ } 2 + 2 \cdot 5$

18. $\frac{3}{4} \underline{\ ?\ } 0.752$

Compute.

19. 0.89×7.5

20. 31.5% of 42

21. $^{-}8 + {}^{+}0.6$

22. $^{-}6 \times {}^{-}\frac{2}{3}$

Write an equation and solve.

23. 15 more than *n* is equal to 41. Find the value of *n*.

24. Alicia's age divided by 3, minus 6, is 18. Find Alicia's age.

25. 4 times a number, increased by 5, is 17.
What is the number?

Solve.

26. Jan worked 4 hours on a job and Ted worked 5 hours.
Together they earned $31.50. How much did each receive?

27. Last week Martin's Newstand sold 325 newspapers, 120 magazines, and 79 paperbacks. This week total sales increased approximately 6%. About how many of the three items combined were sold this week?

28. The average distance of the earth from the sun is 93,000,000 mi.
Express this distance in scientific notation.

29. Write the converse of this statement: If it is raining, then it is spring.

30. Halley's Comet appears every 76 years. How many times will it be seen in 692 years?

Answers

16. ______ (1–2)

17. ______ (3–1)

18. ______ (2–8)

19. ______ (1–6)

20. ______ (7–3)

21. ______ (4–3)

22. ______ (4–6)

23. ______ (3–6)

24. ______ (3–8)

25. ______ (3–8)

26. ______ (6–6)

27. ______ (7–10)

28. ______ (5–3)

29. ______ (5–9)

30. ______ (1–13)

***Next to each item is given the lesson number in the text where the item was taught.**

Profit and Loss*

Name ______________________

Date ______________________

Find the profit or loss to the nearest cent. Use a formula. $P = C \times R$ $L = C \times R$

1. Cost: $80
Rate of Profit: $12\frac{1}{2}\%$

______

2. Cost: $32.64
Rate of Profit: 25%

______

3. Cost: $196
Rate of Loss: $33\frac{1}{3}\%$

______

4. Cost: $624
Rate of Loss: 10%

______

5. Cost: $575.25
Rate of Loss: 3.5%

______

6. Cost: $99.98
Rate of Profit: 3.4%

______

Find the profit or loss to the nearest cent. Use proportion.

7. Cost: $82.95
Rate of Loss: $8\frac{1}{3}\%$

______

8. Cost: $172.40
Rate of Profit: 15%

______

9. Cost: $412.50
Rate of Loss: 8.2%

______

10. Cost: $615.45
Rate of Loss: 4.5%

______

11. Cost: $272.62
Rate of Loss: $6\frac{1}{4}\%$

______

12. Cost: $95.98
Rate of Loss: 2.6%

______

Find the selling price. (Gain: $SP = C + P$; Loss: $SP = C - L$)

13. Cost: $86.50
Gain: $12.90

______

14. Cost: $114.20
Loss: $26.10

______

15. Cost: $320.25
Loss: $4.32

______

16. Cost: $156.40
Gain: $22.15

______

17. Cost: $681.11
Gain: $47.90

______

18. Cost: $587.62
Loss: $37.40

______

Solve.

19. A house was originally priced at $132,515. It was sold at a 12% loss. How much money was lost on the sale? ______________________

20. A florist made a profit of 15% on holiday sales. The flowers cost him $7800. What was the total selling price? ______________________

*Use with Lesson 8-1, text pages 204–205.

Discount and Sale Price*

Name ______________________

Date ______________________

Find the discount and the sale price. $D = LP \times R$ $SP = LP - D$

1. Regular price: $156.99
   $33\frac{1}{3}$% off

   Discount: __________

   Sale price: __________

2. Original price: $24.75
   4% off

   Discount: __________

   Sale price: __________

3. List price: $32
   16% off

   Discount: __________

   Sale price: __________

4. List price: $44.99
   12% discount

   Discount: __________

   Sale price: __________

5. Regular price: $56
   8% discount

   Discount: __________

   Sale price: __________

6. Original price: $145
   10% discount

   Discount: __________

   Sale price: __________

7. Regular price: $14.95
   Rate: 7%

   Discount: __________

   Sale price: __________

8. List price: $25.50
   Rate: 2.5%

   Discount: __________

   Sale price: __________

9. List price: $77.20
   Rate: 16%

   Discount: __________

   Sale price: __________

10. Original price: $95.25
    Rate: 8%

    Discount: __________

    Sale price: __________

11. Regular price: $80
    Rate: 25%

    Discount: __________

    Sale price: __________

12. List price: $45.60
    Rate: 18%

    Discount: __________

    Sale price: __________

13. List price: $186.25
    Rate: 20%

    Discount: __________

    Sale price: __________

14. Regular price: $350
    Rate: 12%

    Discount: __________

    Sale price: __________

15. Original price: $1400
    Rate: 15%

    Discount: __________

    Sale price: __________

***Use with Lesson 8-2, text pages 206–207.**

Rate of Discount and List Price*

Name ______________________

Date ______________________

$$D = LP \times R \text{ or } \frac{D}{LP} = \frac{R \text{ of } D}{R \text{ of } LP}$$

Find the rate of discount.

1. List price: $96.50
Discount: $28.95

Rate: __________

2. List price: $33.60
Discount: $6.72

Rate: __________

3. List price: $124.00
Discount: $17.36

Rate: __________

4. List price: $89.98
Discount: $44.99

Rate: __________

5. List price: $63.18
Discount: $42.12

Rate: __________

6. List price: $453.00
Discount: $67.95

Rate: __________

Find the list price.

7. Discount: $225
Rate: 5%

List price: __________

8. Discount: $5.67
Rate: 9%

List price: __________

9. Discount: $7.56
Rate: 12%

List price: __________

10. Discount: $72.80
Rate: 14%

List price: __________

11. Discount: $12.75
Rate: 15%

List price: __________

12. Discount: $241
Rate: $33\frac{1}{3}$%

List price: __________

Solve.

13. Steve bought a book with a list price of $9.80. He received a discount of 49¢. Find the rate of discount.

____

14. Laura bought a bike on sale for $7.16 off the list price. The discount was 8%. What was the list price?

____

*Use with Lesson 8-3, text pages 208–209.

Sales Tax*

Name ______________________

Date ______________________

$T = MP \times R$ or $\frac{T}{MP} = \frac{R \text{ of } T}{R \text{ of } MP}$

Find the sales tax.

1. Price: $89.45
   Tax: 5%

   Sales tax: ____________

2. Price: $593.80
   Tax: 6%

   Sales tax: ____________

3. Price: $210.70
   Tax: 5%

   Sales tax: ____________

4. Price: $56.49
   Tax: $2\frac{1}{2}$%

   Sales tax: ____________

5. Price: $152.00
   Tax: 4%

   Sales tax: ____________

6. Price: $75.99
   Tax: 6%

   Sales tax: ____________

7. Price: $138.40
   Tax: 7%

   Sales tax: ____________

8. Price: $29.75
   Tax: 3%

   Sales tax: ____________

9. Price: $48.21
   Tax: 5.8%

   Sales tax: ____________

Complete the chart. (Round tax to the nearest cent.)

	Price	Tax	Amount of Tax	Total Cost
10.	$23.95	5%		
11.	$348.60	5%		
12.	$72.20	5%		
13.	$99.50	5%		
14.	$559.25	5%		
15.	$279.85	5%		

Find the total cost including 6% sales tax.

16. Regular price: $4.95
    Discount: 20%

    Total cost: __________

17. Regular price: $12.50
    Discount: 14%

    Total cost: __________

18. Original price: $10.90
    Discount: 10%

    Total cost: __________

***Use with Lesson 8-4, text pages 210–211.**

Commission*

Name ______________________

Date ______________________

$$C = TS \times R \text{ or } \frac{C}{TS} = \frac{R \text{ of } C}{R \text{ of } TS}$$

Find the commission.

1. Total sales: $4800
Rate: 2.5%

Commission: __________

2. Total sales: $2400
Rate: 2%

Commission: __________

3. Total sales: $6000
Rate: 1.5%

Commission: __________

4. Total sales: $8276.20
Rate: 5%

Commission: __________

5. Total sales: $8730
Rate: 3%

Commission: __________

6. Total sales: $548
Rate: 32%

Commission: __________

7. Total sales: $4725
Rate: 3%

Commission: __________

8. Total sales: $976
Rate: 4%

Commission: __________

9. Total sales: $490
Rate: 4%

Commission: __________

10. Total sales: $3550
Rate: 5%

Commission: __________

11. Total sales: $649
Rate: 6%

Commission: __________

12. Total sales: $9632
Rate: 3.5%

Commission: __________

Solve.

13. Mrs. Deno earns a weekly salary of $250 plus a 1.5% commission on all sales she makes. If her sales for one week totaled $7841.56, what was her total income for the week? ______________________

14. During a sale Mr. White sold a piano for $1560. He received a commission of 4.5%. What was the amount of commission? ______________________

15. Nicole is paid $200 a week plus a 6% commission on all sales over $1000. What were here earnings for a week in which she sold $3400 worth of goods? ______________________

*Use with Lesson 8-5, text pages 212–213

Rate of Commission and Total Sales*

Name ______________________

Date ______________________

Find the rate of commission.

1. Commission: $40
   Total sales: $500

   Rate: __________

2. Commission: $120
   Total sales: $2000

   Rate: __________

3. Commission: $1275
   Total sales: $8500

   Rate: __________

4. Commission: $63.75
   Total sales: $1275

   Rate: __________

5. Commission: $3650.75
   Total sales: $73,015

   Rate: __________

6. Commission: $3436
   Total sales: $85,900

   Rate: __________

Find the total sales.

7. Commission: $27.00
   Rate: 12%

   Total sales: __________

8. Commission: $48.00
   Rate: 2%

   Total sales: __________

9. Commission: $162
   Rate: 4%

   Total sales: __________

10. Commission: $139
    Rate: 10%

    Total sales: __________

11. Commission: $250.25
    Rate: 5.5%

    Total sales: __________

12. Commission: $195
    Rate: $3\frac{1}{4}$%

    Total sales: __________

Solve.

13. An agent earned $200 for collecting rents at 4%. How much rent did he collect? ______________________

14. An electrician received $45 for installing a security system in a building. If the system cost $225, what rate of commission did he earn? ______________________

15. Bob works in a music store. One day he earned $40 for selling 5 stereo cassette recorders at $80 each. What was his rate of commission on these sales? ______________________

*Use with Lesson 8-6, text pages 214–215.

Simple Interest*

Name ______________________

Date ______________________

Find the interest.

1. $p = \$1600$
$r = 5\%$
$t = 6$ yr

$I =$ __________

2. $p = \$1260$
$r = 6\%$
$t = 5$ yr

$I =$ __________

3. $p = \$300$
$r = 3\%$
$t = 7$ yr

$I =$ __________

4. $p = \$2400$
$r = 2\frac{1}{2}\%$
$t = 4$ yr

$I =$ __________

5. $p = \$960$
$r = 4\%$
$t = 8$ yr

$I =$ __________

6. $p = \$200$
$r = 6\%$
$t = 2$ yr

$I =$ __________

7. $p = \$150$
$r = 5\%$
$t = 3$ yr

$I =$ __________

8. $p = \$400$
$r = 4\frac{1}{2}\%$
$t = 1$ yr

$I =$ __________

9. $p = \$1260$
$r = 2\frac{1}{4}\%$
$t = 4$ yr

$I =$ __________

Solve.

10. On Dec. 6, 1990, Mr. Dixon borrowed \$3600. The time of the loan was 3 years. How much interest did he pay if the bank charged $14\frac{1}{2}\%$ interest? ______________________

11. Mrs. Helin secured a loan of \$1860 from the bank. How much money will she have to pay back on a 3-year loan at a 6% interest rate? ______________________

12. What will be the mount due on a loan of \$960 at the rate of $6\frac{1}{2}\%$ for 3 years? ______________________

13. Miss Ennis deposited \$450 on June 1, 1990. If the bank paid 5.4% interest, how much did she have in the bank on June 1, 1994? ______________________

14. A club borrowed \$3000 for 2 years at $4\frac{1}{2}\%$ interest. What was the amount due on the loan? ______________________

*Use with Lesson 8-7, text pages 216–217.

Compound Interest*

Name ______________________

Date ______________________

Complete each chart.

1. Find the compound interest on $5000 at 6% for 2 years compounded annually.

Time Period	Principal	Interest
1 year	$5000	
2 years		

2. Find the compound interest on $5000 at 6% for 2 years compounded semiannually.

Time Period	Principal	Interest
$\frac{1}{2}$ year	$5000	
1 year		
$1\frac{1}{2}$ years		
2 years		

Solve.

3. A customer took out a loan of $7500 at 15% interest compounded quarterly for 15 months. How much did he have to pay back? ______________________

4. A family deposited $500 in a savings account to prepare for a trip. If their money earns $12\frac{1}{2}$% interest compounded quarterly, how much will they earn in one year? ______________________

5. Ms. Salta deposited $2000 in a savings account for one year at $2\frac{1}{2}$% simple interest. Mr. Mendoza deposited the same amount in a savings bank which pays $2\frac{1}{2}$% interest compounded semiannually. Who receives the greater interest after one year? ______________________

6. What is the difference in interest on $2500 at 6% compounded annually for 3 years and the same amount invested at 6% simple interest for 3 years? ______________________

*Use with Lesson 8-8, text pages 218–219.

Buying on Credit *

Name ______________________

Date ______________________

Find the monthly finance charge on the unpaid balance. Round to the nearest cent.

1. Bill: $200
   Paid: $75
   Rate: 2%

   Charge:____________

2. Bill: $62.85
   Paid: $12.85
   Rate: $1\frac{1}{2}$%

   Charge:____________

3. Bill: $50.75
   Paid: $15
   Rate: 2.5%

   Charge:____________

4. Bill: $365.50
   Paid: $100
   Rate: $1\frac{1}{2}$%

   Charge:____________

5. Bill: $126.80
   Paid: $40
   Rate: $2\frac{1}{4}$%

   Charge:____________

6. Bill: $288.90
   Paid: $75
   Rate: $1\frac{3}{4}$%

   Charge:____________

Find the total finance charge. Round to the nearest cent.

	Bill	Down Payment	Monthly Payment	Number of Months	Finance Charge
7.	$450	$150	$35	10	
8.	$96	$30	$20	4	
9.	$125	$25	$25	5	
10.	$260	$50	$40	6	
11.	$800	$250	$50	12	
12.	$80	25%	$20	4	
13.	$150	25%	$25	5	
14.	$500	10%	$40	12	
15.	$365	$65	$40	8	
16.	$76	$26	$10	6	

Solve:

17. A real estate broker used her credit card to buy an answering machine for $180. She paid $40 a month for two months and the balance the third month. If there was a $1\frac{1}{2}$% finance charge after the first month, how much did she actully pay for the machine?

____

*Use with Lesson 8-9, text pages 220–221.

Probability of a Single Event*

Name ______________________

Date ______________________

Find the probability.

Experiment A: Toss a cube with sides marked 1, 2, 3, 4, 5, and 6.

1. $P(3)$ __________ **2.** P(odd number) __________ **3.** P(multiple of 3) __________

4. $P(4)$ __________ **5.** P(2-digit number) __________ **6.** P(1-digit number) __________

7. $P(< 5)$ __________ **8.** P(2 or 3) __________ **9.** $P(\geq 6)$ __________

Experiment B: Spin the dial.

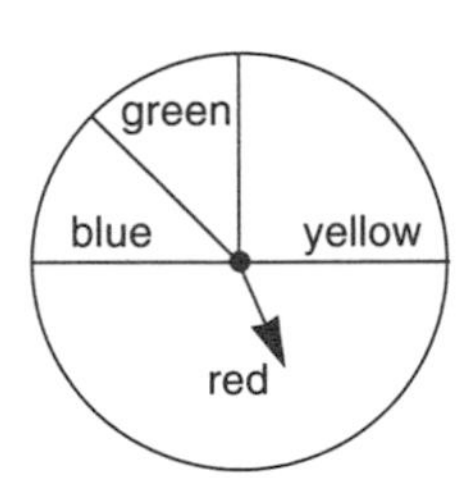

10. P(red) __________ **11.** P(green) __________

12. P(yellow) __________ **13.** P(blue) __________

14. P(yellow or red) __________ **15.** P(black) __________

Experiment C: Choose a card from an envelope containing 10 cards numbered 1 to 10.

16. $P(5)$ __________ **17.** P(even) __________ **18.** P(greater than 2) __________

19. P(multiple of 4) __________ **20.** P(divisible by 5) __________ **21.** P(less than 1) __________

22. P(5 or 7) __________ **23.** P(odd) __________ **24.** P(less than 9) __________

25. Project: Complete the chart by flipping a coin for the number of flips given. Each time find the probability of getting heads.

Number of Flips	Number of heads	Number of tails	*P* (heads)
10			
20			
30			
40			
50			

*Use with Lesson 9-1, text pages 240–241.

Independent, Dependent and Compound Events*

Name ______________________________
Date ______________________________

Find the probability.

Experiment A: Spin the dial twice.

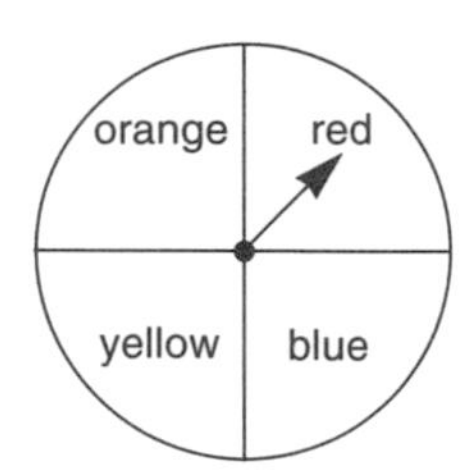

1. *P* (red, blue) __________
2. *P* (orange, yellow) __________
3. *P* (blue, red) __________
4. *P* (red, blue or orange) __________

Experiment B: Spin the dial twice.

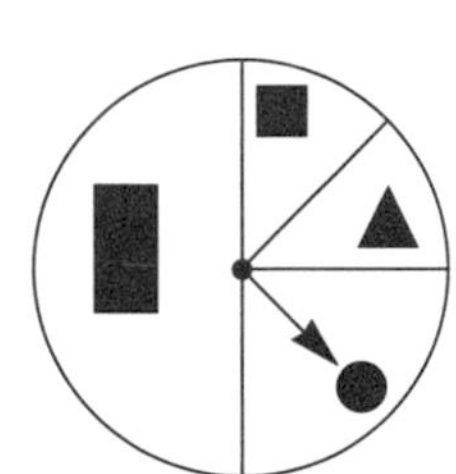

5. *P* (■, ●) __________
6. *P* (▲, ●) __________
7. *P* (▮, ▲) __________
8. *P* (■, ▲) __________
9. *P* (●, ▮) __________
10. *P* (▮, ● or ▲) __________

Find each probability. Complete the chart.

Experiment C: Pick two cards.

1 2 3 4 5 6 7 8

		Replacing first card	Not replacing first card
11.	*P* (1, 2)		
13.	*P* (4, 8)		
15.	*P* (odd, even)		
17.	*P* (even, even)		
19.	*P* (2, odd)		

		Replacing first card	Not replacing first card
12.	*P* (3, even)		
14.	*P* (4, prime)		
16.	*P* (3, prime)		
18.	*P* (4, even)		
20.	*P* (1, 10)		

*Use with Lesson 9-2, text pages 242–243.

Permutations*

Name ______________________

Date ______________________

Find the value of each factorial expression.

1. $6!$ ______

2. $3!$ ______

3. $8!$ ______

4. $4! \cdot 2!$ ______

5. $(5 \cdot 2)!$ ______

6. $6! + 3!$ ______

7. $8! - 3!$ ______

8. $9! - 6!$ ______

9. $7! \div (6 - 2)!$ ______

Solve.

10. There are 10 entries in the local dog show. Three ribbons will be awarded. In how many ways can the ribbons be awarded?

11. The swimming relay team has 4 members. How many ways can the coach set the order in which they will swim?

12. You have 7 cards. In how many ways can they be arranged?

13. In how many ways can the letters M, E, D, A, R be arranged? What is the probability that a random arrangement of these letters will form the word DREAM?

14. Four skydivers were entered in a target-landing contest. How many different orders of finish are possible?

*Use with Lesson 9-3, text pages 244–245.

Random Samples*

Name ______________________

Date ______________________

Complete each chart. Then find the estimates.

Question asked of 60 people: What is your favorite TV station for sports?

	Responses	Number	% of Sample
1.	Channel 3 𝍸 𝍸 𝍸		
2.	Channel 6 𝍸 𝍸 𝍸 𝍸 𝍸		
3.	Channel 9 𝍸 𝍸 𝍸 𝍸		

Estimate the number of people in a population of 240
who would prefer these stations:

4. Channel 3 ____________ **5.** Channel 6 ____________ **6.** Channel 9 ____________

Estimate the number of people in a population of 600
who would prefer these stations:

7. Channel 3 ____________ **8.** Channel 6 ____________ **9.** Channel 9 ____________

Question asked of ________ high school seniors: What is the maximum level of schooling you plan to complete?

	Responses	Number	% of Sample
10.	High School 𝍸 I		
11.	Junior College 𝍸 𝍸 𝍸 𝍸 II		
12.	Trade School 𝍸 𝍸 IIII		
13.	College 𝍸 𝍸 𝍸 𝍸 𝍸 𝍸 I		
14.	Post Graduate 𝍸 𝍸 𝍸 𝍸 II		

Estimate the number of high school seniors in a
class of 300 who expect to complete the following:

15. High School ________ **16.** Junior College ________ **17.** Trade School ________

18. College ________ **19.** Post Graduate ________

*Use with Lesson 9-4, text pages 246–247.

Pictographs*

Name ______________________

Date ______________________

Answer these questions about the pictographs below.

Attendance at Sports Events

= 100 fans

Football	
Basketball	
Hockey	
Swimming	
Soccer	

1. About how many fans attend basketball games? ____________
2. About how many fans attend swimming meets? ____________
3. What two sports have almost the same attendance?

   ____

4. What is the difference in attendance at football games and hockey games?

   ____

Branch Libraries Circulations

January – February = 1000 Books

Drexel	
Homville	
Moreland	
Olney	
Pawmont	
Bayside	

5. What is the circulation at the busiest branch? ____________
6. What is the circulation at Bayside? ____________
7. What is the difference in circulation between Pawmont and Moreland?

   ____

8. What is the total circulation at these six branches?

   ____

9. **Construct a pictograph for this chart.**

Favorite Ice-Cream Flavors

Flavors	Number
Chocolate	25 people
Vanilla	22 people
Mint Chip	20 people
Strawberry	12 people
Butterscotch	16 people

Pictograph

*Use with Lesson 9-5, text page 248.

Bar Graphs*

Name ______________________________

Date ______________________________

Complete this frequency table and the bar graph for it.

1.

Baskets Scored in Free-Throw Contest		
Contestant	**Tally**	**Number**
Rob	𝍸 𝍸 I	
Kate	𝍸 𝍸 III	
Agnes	𝍸 III	
Scott	𝍸 𝍸	
Manuel	𝍸 II	
Chen	𝍸 𝍸	

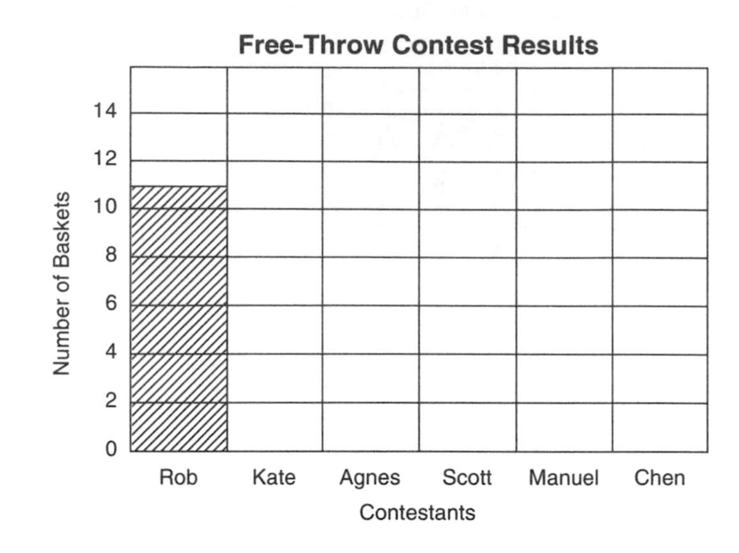

Construct a bar graph for each chart.

2.

Totals in Fund-Raiser	
Grade	**Amount**
Grade 4	$100
Grade 5	$200
Grade 6	$150
Grade 7	$225
Grade 8	$300
Grade 9	$250

3.

Oceans of the Earth	
Ocean	**Area**
Atlantic	82 million km^2
Pacific	166 million km^2
Indian	66 million km^2
Arctic	14 million km^2

*Use with Lesson 9-6, text page 249.

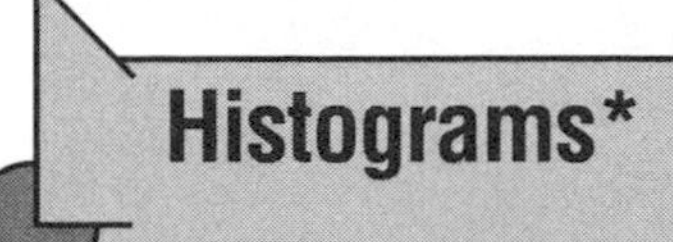

Histograms*

Name ______________________

Date ______________________

Answer the questions about the histograms.

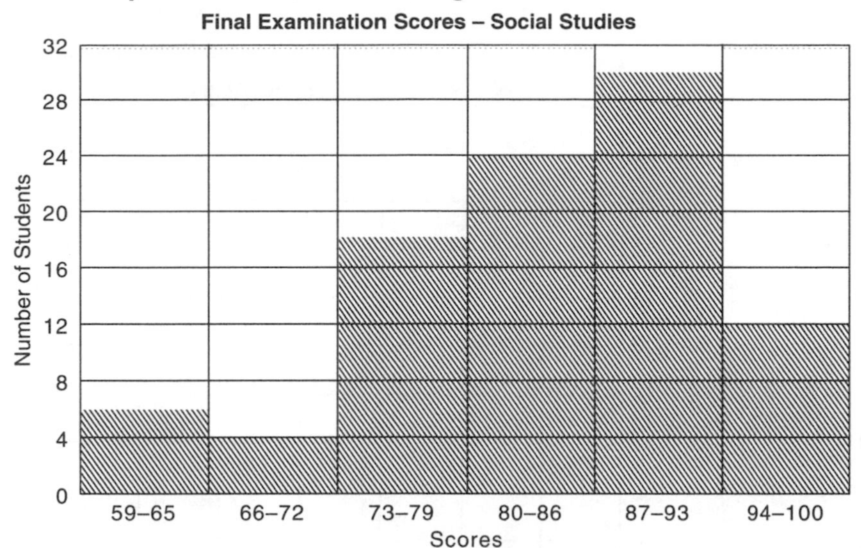

Grade Chart
A = 94–100
B = 87–93
C = 80–86
D = 73–79
F = below 73

1. How many students scored between 87 and 93 on the exam? ______________________
2. Which group is larger and by how much: those who scored 80 or more or those who scored less than 80? ______________________
3. How many students took the examination? ______________________
4. Letter grades were given according to the grade chart shown above. How many students received A or B? ______________________
5. How many students received C or D? ______________________
6. How many students received F? ______________________

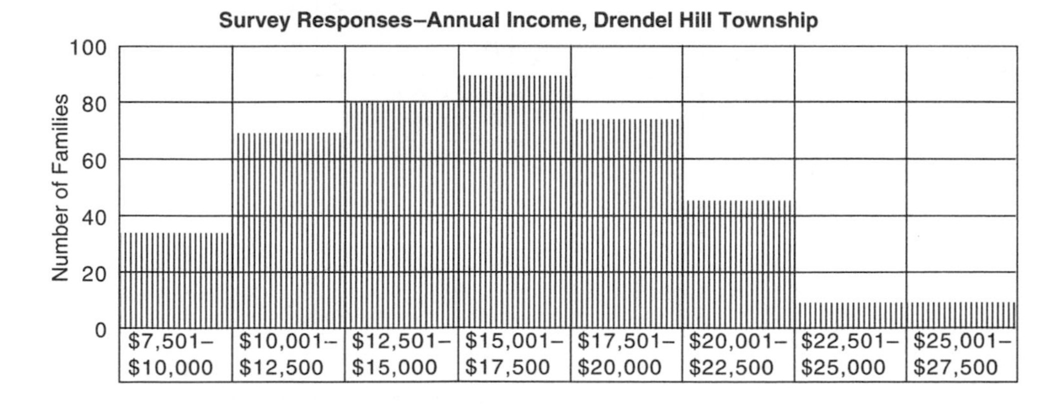

7. About how many families have an income between $10,001 and $12,500? ______________________
8. About how many families have an income between $12,501 and $20,000? ______________________
9. About how many families answered the survey? ______________________

*Use with Lesson 9-7, text page 250.

Line Graphs*

Name ____________________

Date ____________________

1. Complete the line graph for the information in the chart.

Week's High Temperatures	
Day	**Temperature, °C**
Sunday	20°
Monday	22°
Tuesday	26°
Wednesday	30°
Thursday	28°
Friday	28°
Saturday	32°

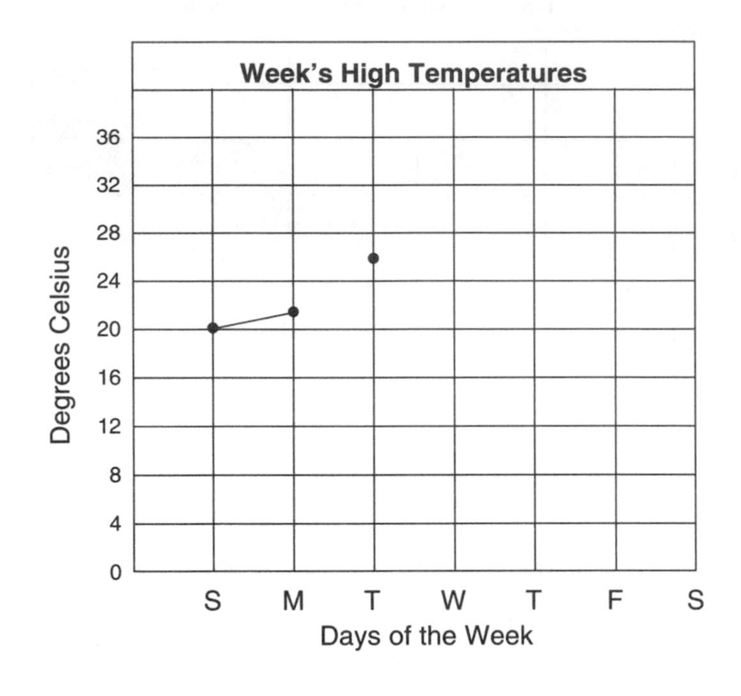

2. Complete the double line graph.

Monthly High Temperatures		
	Temperature, °F	
Month	Los Angeles	San Francisco
Jan	65°	56°
Feb	66°	59°
Mar	69°	61°
Apr	71°	63°
May	74°	65°
Jun	77°	69°
July	83°	69°
Aug	84°	70°
Sept	82°	72°
Oct	77°	69°
Nov	73°	64°
Dec	67°	57°

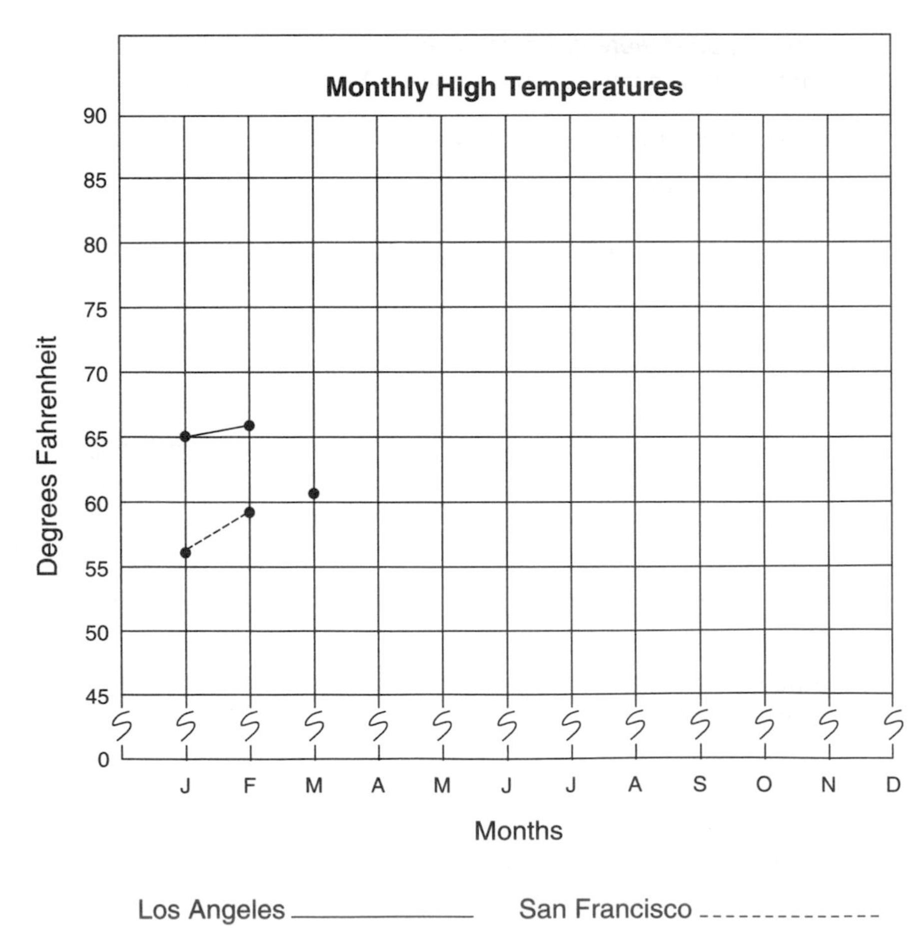

*Use with Lessons 9-8, 9-9, text pages 251–253.

Circle Graphs*

Name ______________________

Date ______________________

How many degrees in a circle graph are needed to show each?

1. 5 out of 20 people ______________
2. 4 out of 40 doctors ______________
3. 8 out of 24 hours ______________
4. 150 out of 200 books ______________
5. 20 out of 30 days ______________
6. 45 out of 54 children ______________
7. 60 out of 100 dollars ______________
8. 21 out of 56 votes ______________

Find each percent.

9. John spends his allowance as follows:

   Books: $3.75 __________ %

   Lunches: $2.50 __________ %

   Records: $2.50 __________ %

   Savings: $1.25 __________ %

10. Kate spends her leisure time each week as follows:

   Reading: 8 hours __________ %

   Jogging: 8 hours __________ %

   Movies or TV: 4 hours __________ %

   Others: 4 hours __________ %

Complete the chart. Then construct a circle graph for it.

11.

Major Uses of Energy in the U.S.		
Use	**Percent**	**Degrees**
Heat for industry	35	
Transportation	24	
Heat for buildings	18	
Electricity for industry	12	
Water heating	4	
Air conditioning	2.3	
Refrigeration	2.1	
Lighting	1.3	
Cooking	1.3	

*Use with Lesson 9-10, text pages 254–255.

Central Tendency*

Name ______________________

Date ______________________

Find the range, mean, median, and mode for each set of data.

1. Savings for the past 10 months:
   $25, $56, $97, $38, $50
   $46, $49, $52, $61, $75

   range: ______ **median:** ______

   mean: ______ **mode:** ________

2. Expenses for the past 10 months:
   $91, $107, $82, $98, $91
   $49, $86, $55, $51, $66

   range: ______ **median:** ______

   mean: ______ **mode:** ________

3. Number of people on 8 tours:
   60, 42, 48, 35
   51, 28, 39, 64

   range: ______ **median:** ______

   mean: ______ **mode:** ________

4. Test scores for 8 students:
   72, 79, 91, 84
   96, 88, 81, 70

   range: ______ **median:** ______

   mean: ______ **mode:** ________

Solve.

5. The softball team won the league batting title. The individual averages were 0.325, 0.380, 0.298, 0.308, 0.312, 0.300, 0.292, 0.288, 0.316. Compute the team average, the range, and the median.

   average: ______ **range:** ______ **median:** ______

6. At summer camp, a record number of campers were enrolled in 8 fields of activity: 38, 47, 26, 52, 59, 65, 53, 32. Compute the average number, the range, and the median.

   average: ______ **range:** ______ **median:** ______

7. A cyclist kept a record of the distance she biked each day for a week. The record was as follows: 14.5 mi, 9.6 mi, 5.6 mi, 10.7 mi, 15.1 mi, 4.3 mi, and 5.6 mi. Find the mean and the mode.

   mean: ______ **mode:** ______

8. The track team's times in the 200-meter dash were 26.1, 27.5, 27.2, 26.2, 25.0, 24.3, and 24.9. How many runners' times were above the team average?

   above average: ______

***Use with Lesson 9-11, text pages 256–257.**

Geometric Concepts*

Name ______________________

Date ______________________

Match each symbol with the corresponding figure.

1. _____ $\overrightarrow{CT}$
2. _____ $\overrightarrow{SR}$
3. _____ $\overleftrightarrow{QS}$
4. _____ PQR
5. _____ $\angle RWM$
6. _____ $\angle MRW$
7. _____ $\overline{TP}$
8. _____ M

a. C T

b. Q S

c. 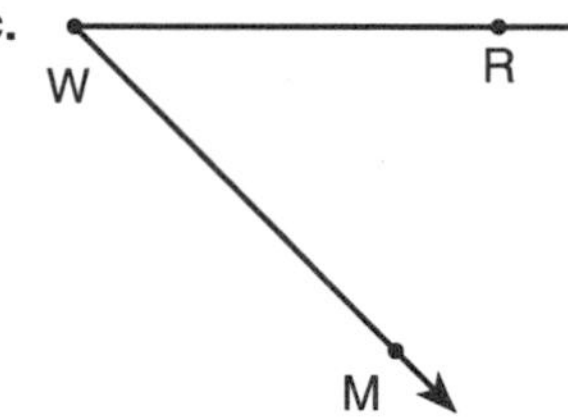

d. 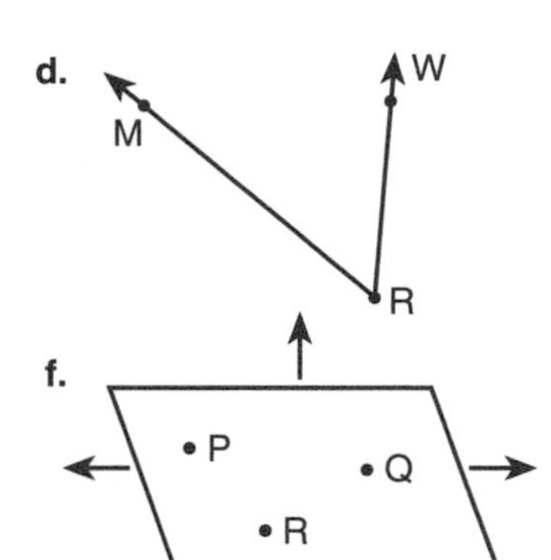

e. 

f.

g. 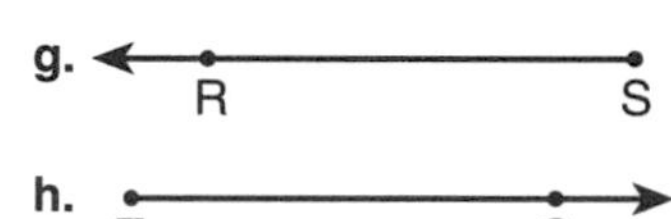

h.

i. M

Draw the figure.

9. $\overrightarrow{AB}$	10. $\angle BML$	11. J
12. $\overline{MN}$	13. $\overleftrightarrow{RT}$	14. $\angle TRS$

Write "True" or "False."

15. A line can contain a point. ______________________
16. A ray contains a line. ______________________
17. $\overleftrightarrow{AB}$ and $\overline{AB}$ name the same set of points. ______________________
18. The vertex of $\angle CBA$ is the point C. ______________________
19. $\overrightarrow{BA}$ and $\overrightarrow{AB}$ are different rays. ______________________
20. $\overrightarrow{RS}$ is one of the sides of $\angle SRA$. ______________________
21. $\angle TMN$ contains $\overrightarrow{TM}$. ______________________
22. A side of an angle contains a segment. ______________________

*Use with Lesson 10-1, text pages 272–273.

Measuring and Classifying Angles*

Name ____________________

Date ____________________

Measure each angle. Then label it as right, straight, obtuse, or acute.

1.

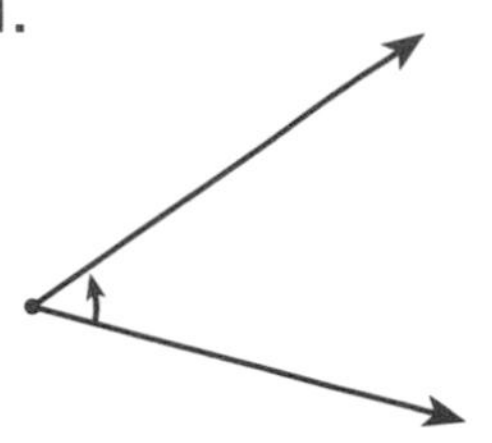

Measure: __________

Kind: ____________

2.

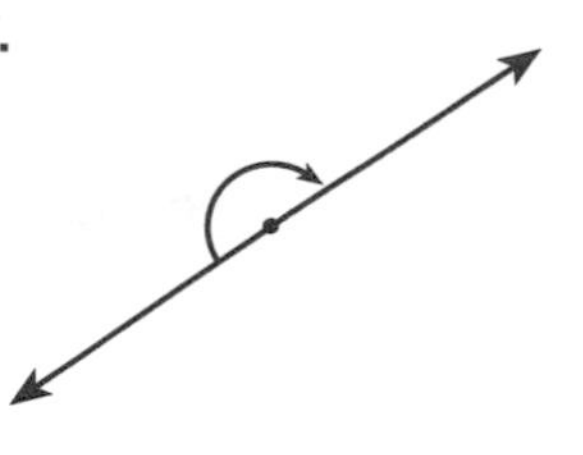

Measure: __________

Kind: ____________

3. **4.**

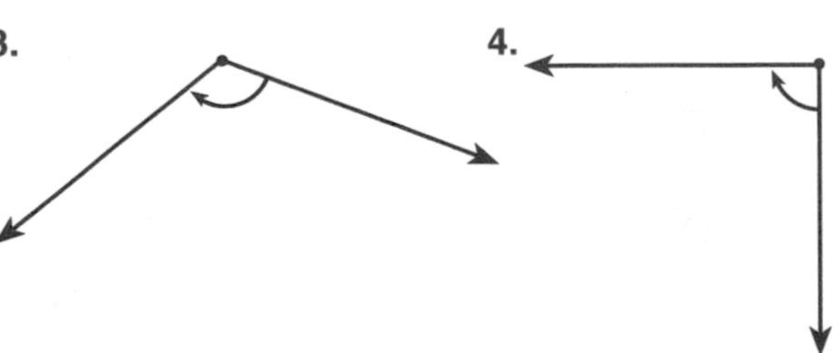

3. Measure: __________

Kind: ____________

4. Measure: __________

Kind: ____________

Draw an angle whose measure is:

5. 100° **6.** 30° **7.** 120°

Complete each chart.

	Measure of $\angle A$	Complement
8.	75°	
9.	88°	
10.	47°	
11.	13°	

	Measure of $\angle B$	Supplement
12.	144°	
13.	14°	
14.	51°	
15.	92°	

Circle the correct answer.

16. If two angles are congruent and complementary, they each measure: **a.** 90° **b.** 45° **c.** 30°

17. Another name for $\angle BON$ is: **a.** $\angle NOB$ **b.** $\angle OBN$ **c.** $\angle NBO$

18. An acute angle has a supplement that is: **a.** acute **b.** obtuse **c.** right

*Use with Lesson 10-2, text pages 274–275.

Constructions with Angles*

Name ______________________

Date ______________________

Using a compass and straightedge, construct an angle congruent to each.

1.

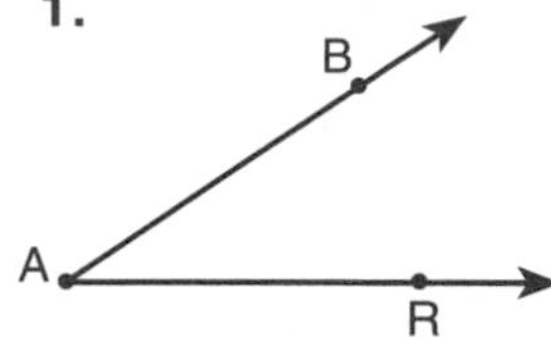

2.

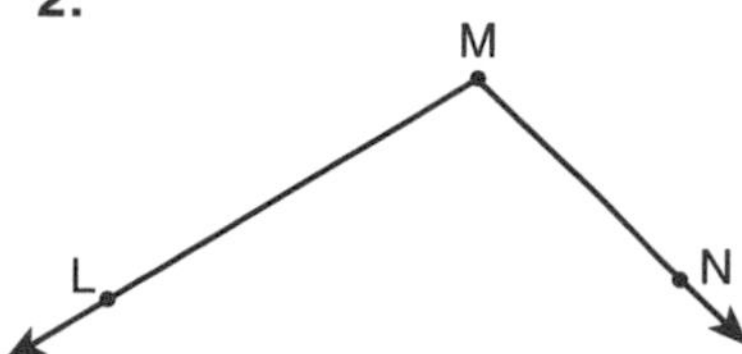

3.

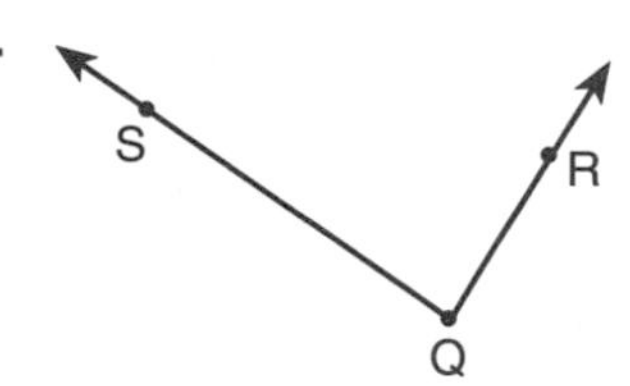

Bisect each angle, using a compass and straightedge.

4.

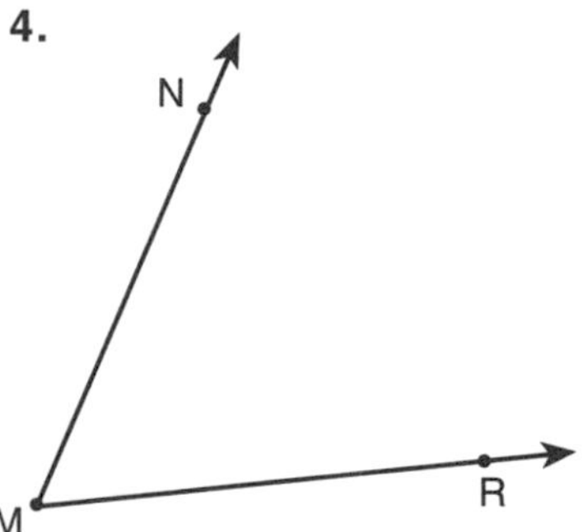

5.

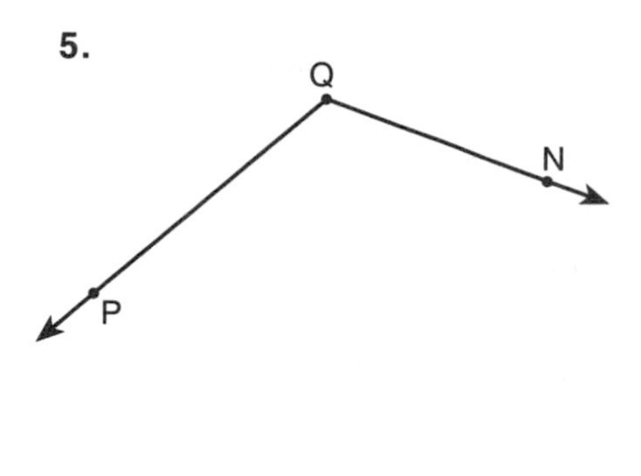

6.

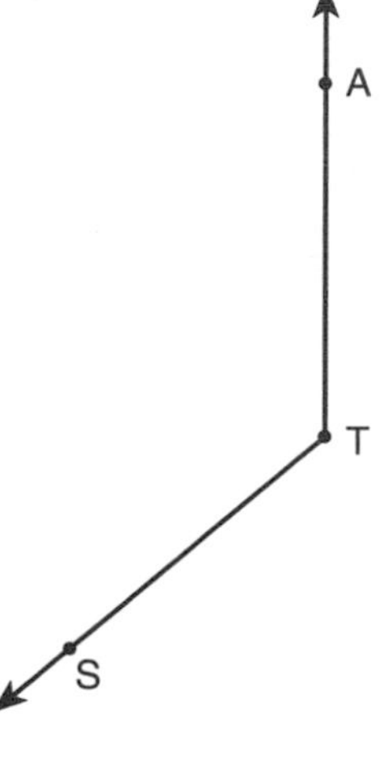

7. Follow the directions.

a. Construct an angle congruent to $\angle PNR$ having $\overrightarrow{NR}$ as one side. Label it $\angle RNS$.

b. Bisect the two congruent angles.

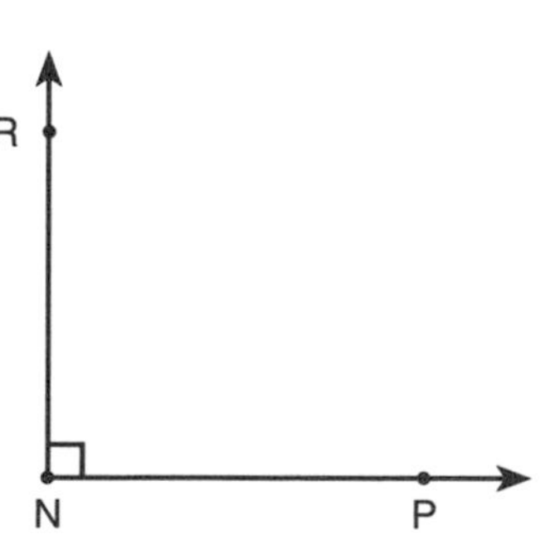

8. Measure each of the four angles formed. What do you find?

____

Parallel, Intersecting, and Perpendicular Lines*

Name ______________________

Date ______________________

Use the figure at the right to answer exercises 1-5.

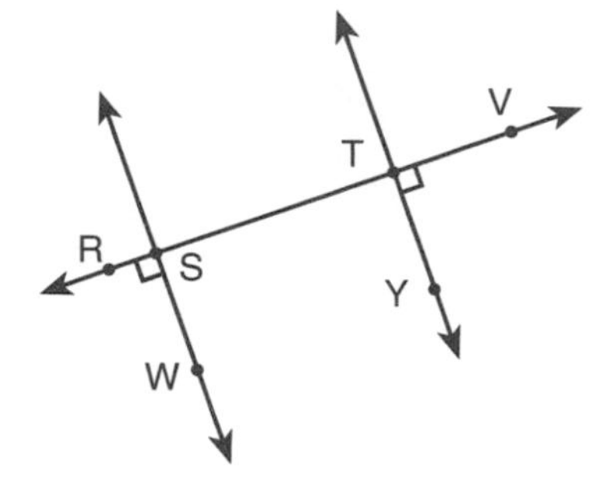

1. Name a pair of parallel lines. ______________________
2. Name a pair of perpendicular lines. ______________________
3. What is the measure of $\angle VTY$? ______________________
4. How many right angles are shown? ______________________
5. What is the measure of $\angle TSW$? ______________________

Use the figure at the left to answer exercises 6-10.

6. Name all pairs of vertical angles. ______________________
7. Name all pairs of adjacent angles. ______________________

   ____
8. Name all pairs of alternate interior angles. ______________________
9. Name all pairs of corresponding angles. ______________________

   ____
10. Name all pairs of alternate exterior angles. ______________________

    ____

Use the figure at the right to answer exercises 11-16.

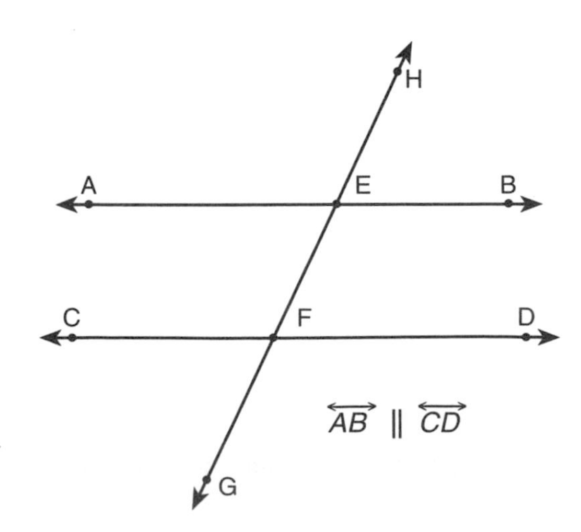

11. Name a pair of parallel lines. ____________
12. Name a transversal. ____________
13. Name the corresponding angle to $\angle AEH$. ____________
14. Name the alternate interior angle to $\angle BEF$. ____________
15. Name the alternate exterior angle to $\angle GFD$. ____________
16. Name two angles adjacent to $\angle EFD$.____________

*Use with Lesson 10-4, text pages 278–279.

Classifying Triangles*

Name ______________________________

Date ______________________________

Supply the missing angle measure for each triangle. Then classify the triangle.

	∠*a*	∠*b*	∠*c*	Type of Triangle
1.	40°	70°		
2.		25°	90°	
3.	72°		47°	
4.	20°	90°		
5.	12°	80°		
6.	15°	75°		
7.	42°	66°		
8.	69°		51°	
9.	100°	15°		
10.		50°	60°	
11.	60°		60°	
12.	40°	100°		
13.	90°	45°		

Answer each question about triangles. Write each letter of the answer on a line. Use the letters in the boxes to answer the riddle below.

14. A triangle with no congruent sides. ___ ___ ___ ___ ___ ☐ ___

15. A triangle with two congruent sides. ☐ ___ ___ ___ ___ ___ ___ ___ ___

16. A triangle with three congruent sides. ___ ___ ___ ___ ___ ___ ☐ ___ ☐ ___ ___

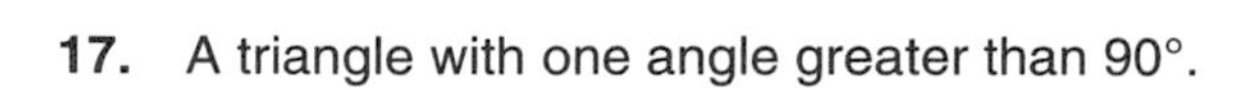

17. A triangle with one angle greater than 90°. ☐ ___ ___ ___ ___ ___

18. A triangle formed by 30°, 60°, 90° angles. ___ ___ ☐ ☐ ___

What did the triangle say to the green light? ☐ ☐ ☐ ☐ ☐ ☐ ☐ !

16 15 18 18 16 17 14

*Use with Lesson 10-5, text pages 280–281.

Congruent Triangles*

Name ______________________

Date ______________________

Complete exercises 1-6 for these congruent triangles.

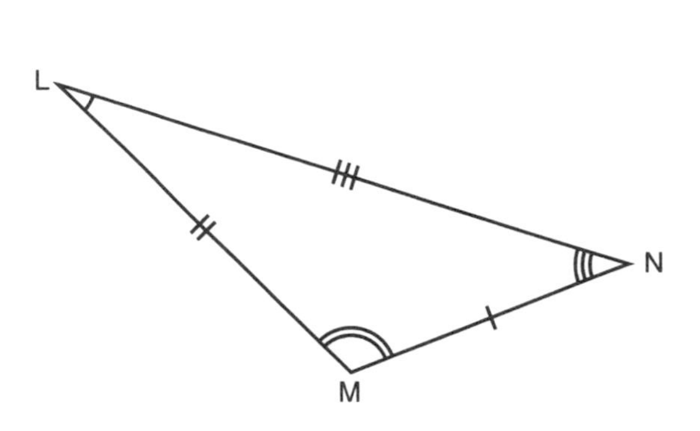

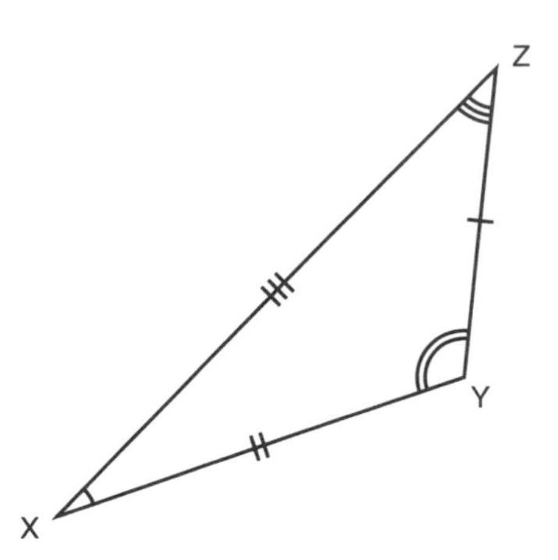

1. $\overline{LN} \cong$ ________
2. $\overline{XY} \cong$ ________
3. $\overline{MN} \cong$ ________
4. $\angle L \cong$ ________
5. $\angle Y \cong$ ________
6. $\angle Z \cong$ ________

Tell which rule, SSS, ASA, or SAS, states why the pairs of triangles are congruent.

7.

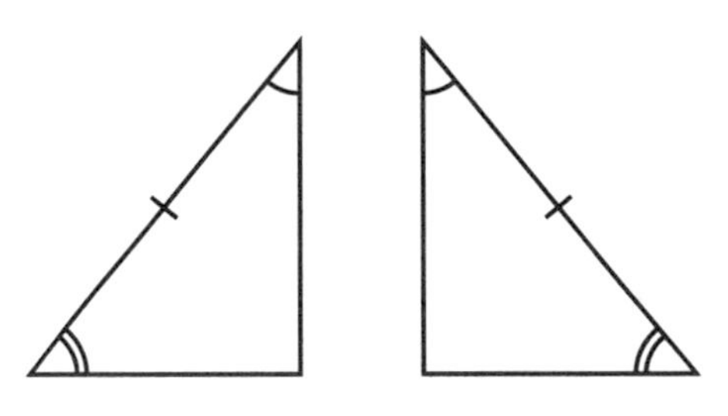

____

8.

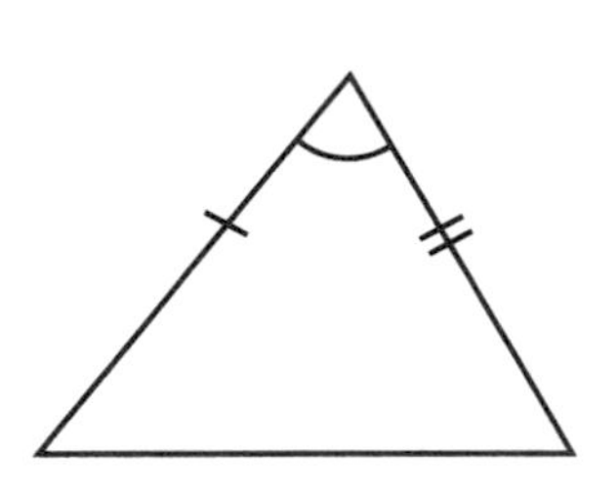

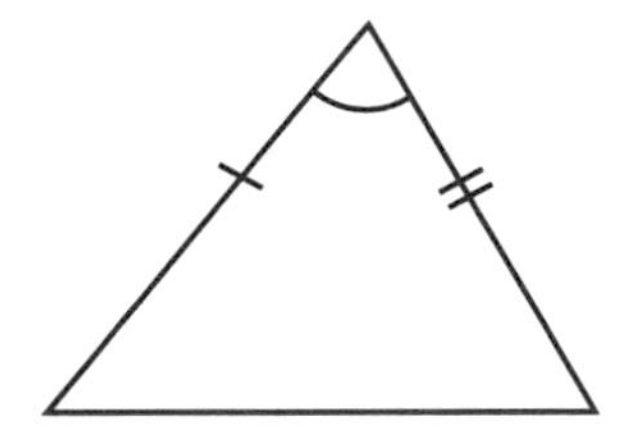

____

9.

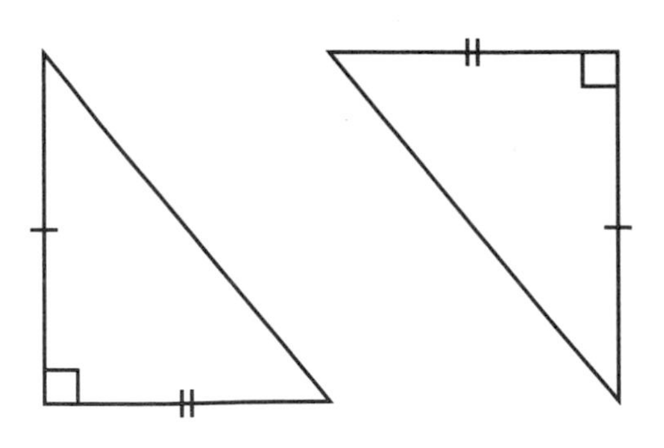

____

10.

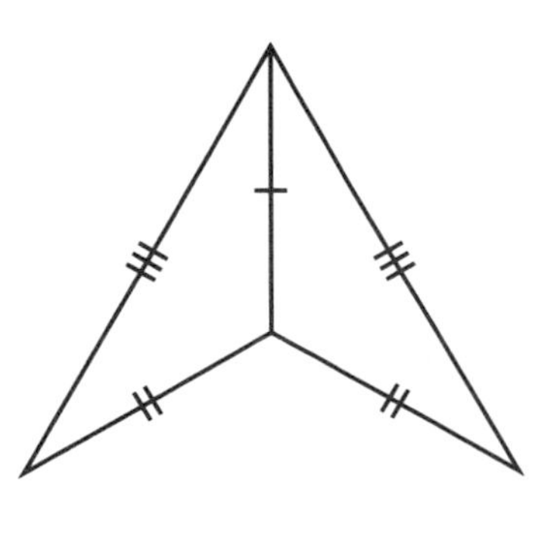

____

*Use with Lesson 10-6, text pages 282–283.

Customary Units*

Name ______________________

Date ______________________

Length	Weight	Capacity
1 ft = 12 in.	1 lb = 16 oz	1 c = 8 fl oz
1 yd = 3 ft = 36 in.	1 T = 2000 lb	1 pt = 16 fl oz
1 mi = 1760 yd = 5280 ft		1 qt = 2 pt = 32 fl oz
		1 gal = 4 qt = 8 pt

Complete.

1. 12 qt = ______ gal

2. 3 yd 3 ft = ______ in.

3. 24 oz = ______ lb

4. $4\frac{1}{4}$ ft = ______ in.

5. $3\frac{1}{2}$ gal = ______ qt

6. 7′ 2″ = ______ in.

7. 3 T = ______ lb

8. 114 oz = ______ lb

9. 4 mi = ______ yd

10. 2 qt 2 pt = ______ fl oz

Compute.

11. 3 lb 9 oz
+ 6 lb 7 oz

12. 1 mi 200 ft
− 780 ft

13. 6 gal 1 qt
− 3 gal 3 qt

14. 3 c 8 fl oz
− 10 fl oz

15. 4 gal 2 qt 1 pt
+ 3 gal 3 qt 1 pt

16. 6 yd 2 ft 5 in.
+ 1 yd 3 ft 4 in.

17. 2(4 qt 1 pt) = __________

18. (75 yd 16 in.) ÷ 4 = __________

Solve.

19. Super-Glo floor wax is sold for $2.89 per quart. How much will 6 gal cost? __________

20. A mechanic has two 1-gal containers for oil. One container is full and the other is $\frac{1}{4}$ full. How many quarts of oil is this? __________

*Use with Lesson 10-7, text pages 284–285.

Metric Units of Length*

Name ______________________

Date ______________________

Complete with km, m, cm, or mm to make reasonable statements.

1. A pencil is approximately 16 _______ long.
2. A book may be 21 _______ thick.
3. An adult's height may be 1.87 _______.
4. The distance Mary walks in 1 hour is 3.5 _______.
5. A jaguar is approximately 75 _______ high at the shoulders.
6. Earth is approximately 152 000 000 _______ from the sun.

Complete.

7. 84 cm = _______ dm
8. 31.6 dam = _______ km
9. 807 dm = _______ km
10. 8 km = _______ hm
11. 3.6 cm = _______ mm
12. 31.6 m = _______ dm
13. 240 mm = _______ m
14. 0.826 m = _______ hm
15. 240 dam = _______ hm
16. 27 m = _______ mm
17. 0.24 dam = _______ dm
18. 20 dam = _______ mm
19. 4000 hm = _______ km
20. 2300 mm = _______ m
21. 0.5 m = _______ km
22. 8000 mm = _______ km
23. 98.1 km = _______ m
24. 1900 mm = _______ dam
25. 2.6 cm = _______ m
26. 27.5 hm = _______ km
27. 300 cm = _______ hm
28. 4.2 cm = _______ mm
29. 0.91 m = _______ dam
30. 0.4 hm = _______ dm
31. 7.2 dm = _______ dam
32. 6200 mm = _______ m
33. 6 km = _______ cm
34. 31 m = _______ cm
35. 810 cm = _______ mm
36. 900 m = _______ hm

Express each measure as a single unit.

37. 8 m 26 cm = _______ cm
38. 9 m 18 mm = _______ m
39. 16 m 47 mm = _______ m
40. 4 cm 3 mm = _______ mm
41. 3 km 2 m = _______ km
42. 8 km 6 hm = _______ m
43. 80 km 5 m = _______ m
44. 6 cm 9 mm = _______ cm
45. 6 km 19 m = _______ m
46. 3 km 6 hm = _______ hm
47. 3 m 46 cm = _______ m
48. 18 hm 7 dm = _______ km

*Use with Lesson 10-8, text pages 286–287.

Changing Metric Units*

Name ______________________

Date ______________________

Complete this chart so that each row shows equivalent measurements.

	kilometer	hectometer	dekameter	meter	decimeter	centimeter	millimeter
1.	0.005	0.05		5	50		5000
2.	0.072	0.72	7.2			7200	
3.	0.0406		4.06			4060	
4.		0.0066		0.66			660
5.	6.8		680			680 000	
6.		0.03			30		3000

Complete. Use the shortcut.

7. 15 m = _______ cm

8. 9 km = _______ m

9. 25 m = _______ mm

10. 4.5 km = _______ dm

11. 6.2 km = _______ dam

12. 525 m = _______ cm

13. 0.46 dam = _______ km

14. 0.82 hm = _______ dm

15. 0.003 dm = _______ mm

Compare. Write <, =, or >.

16. 525 cm _______ 5 m

17. 0.62 hm _______ 62 m

18. 7.4 m _______ 70 dm

19. 52.1 km _______ 521 hm

20. 221 cm _______ 20 hm

21. 0.7 dam _______ 28 mm

Solve.

22. A rug measures 12 m 20 cm. What will it cost to place a border around the rug at $11.50 per meter? _______________

23. The distance from the principal's office to the library is 24 m 7 dm, and the distance from the library to the math workshop room is 24 m 56 cm. Which is the greater distance? How much greater? _______________

*Use with Lesson 10-9, text pages 288–289.

Precision in Measurement*

Name ______________________

Date ______________________

Circle the more precise measurement.

1. 2.5 cm or 0.3 m
2. 19 cm or 18.6 mm
3. 48 mm or 4.81 cm
4. 713 m or 0.71 km
5. 0.6 m or 6.1 dm
6. 85 dm or 9 m
7. 7.3 km or 7314 m
8. 6 hm or 62 m
9. 142 cm or 1.4 m
10. 8.2 mm or 0.8 cm

Complete the chart.

	Measure	Unit	GPE	Range of Measure
11.	16.1 cm			
12.	83 mm			
13.	7 dm			
14.	81 m			
15.	2.7 m			
16.	3 km			
17.	9.6 dam			
18.	0.04 mm			
19.	18 cm			

Measure items around your classroom or your house. Vary the units of measurement you choose. Complete the following chart.

	Item	Measure	Unit	GPE	Range of Measure
20.	Your height				
21.	A glass				
22.	Refrigerator				
23.					
24.					

*Use with Lesson 10-10, text pages 290–291.

Nonroutine Problems*

Name ______________________

Date ______________________

Solve.

1. Ms. Alvarez travels 5 mi east, 6 mi south, 4 mi west, and 6 mi north. How far from the starting point is she?

2. The National Monuments and the White House are part of the National Park System. The White House grounds contain about 18 acres. The acreage of the National Monuments is about 2.69×10^5 times as great. About how many acres are occupied by the National Monuments?

3. Two towns are considering building a walkway over the river that separates them. They poll their citizens to find out how they feel about the project. In town A 360 of the 580 people polled like the idea. In Town B 270 of the 690 people polled like the idea. What percent in each town like the plan? What percent of both towns together like the plan?

4. It costs $13,640 daily to produce 150,000 copies of a daily newspaper. If the circulation remains the same but the prices increase 7% this year, what will be the daily production cost of the newspaper? What is the daily production cost of one newspaper?

5. Students at Progressive University can study mathematics with three different professors. In a sample group of 30 students, 12 study with professor A and 9 with professor B. Of professor A's students, $\frac{1}{3}$ also study with professor B and another $\frac{1}{3}$ study with professor C. Of professor B's students, $\frac{2}{3}$ also study with professor C. Three students study with all of them and four students study with none of them. How many students study with professor C?

*Use with Chapters 5-10.

Nonroutine Problems*
(con't)

Name ______________________________

Date ______________________________

6. Adult meals at a club benefit luncheon sell for $4 each. Is $135 enough to pay for the meals for 34 company executives? Explain.

7. Two students are playing a game of Hopping Squares. There are 29 numbered squares in a straight line separating them. Each player makes a prescribed number of moves. Player A starts at square 1 and moves forward 3 squares and back 1. Player B starts at square 29 and moves forward 5 squares and back 2. If they continue moving toward one another in this way, in which square will they meet?

8. When a large collection of pictures of world leaders is counted by 2's, 3's, 4's, 5's, or 6's, one picture always remains. When counted by 7's, there is no remainder. What is the fewest number of pictures that can be in the collection?

9. One of the angles of a right triangle is 2 more than 3 times the other. What is the measure of each angle?

10. The ratio of the length to width of a rectangle is 3:2. The perimeter is 70 cm. What are the dimensions of the rectangle?

*Use for Chapters 5-10.

Perimeter*

Name ______________________

Date ______________________

Find the perimeter.

1.

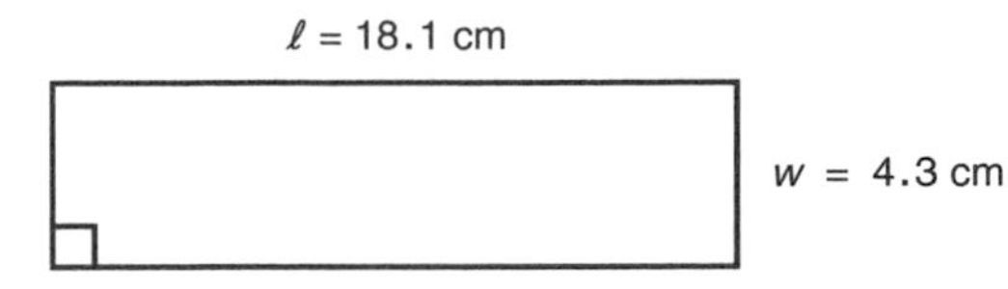

P = ________

2.

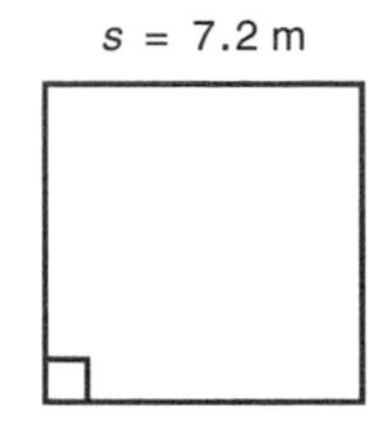

P = ________

3.

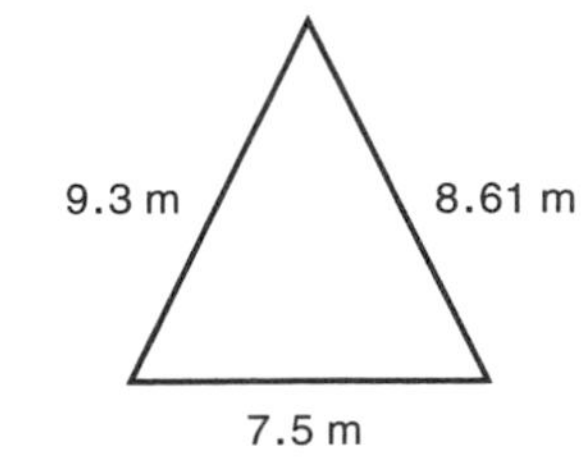

P = ________

4.

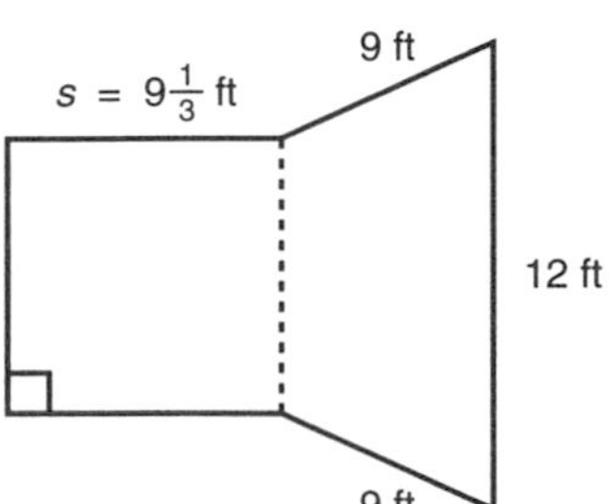

P = ________

5.

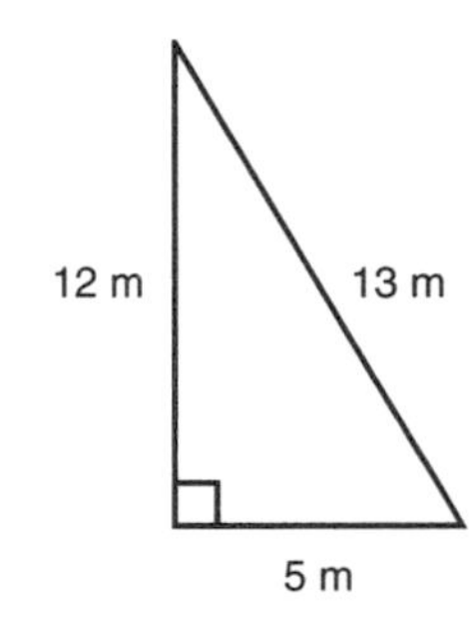

P = ________

6.

18.3 cm
12 cm
13.4 cm
8 cm
3 cm
7.1 cm
8 cm
2.3 cm

P = ________

Draw each figure and solve.

7. Find the perimeter in centimeters of an equilateral triangle measuring 84 mm on a side. ______________________

8. Determine the perimeter of a regular pentagon 3 m 49 cm on a side. ______________________

9. Part of the edging around a regular decagon having a perimeter of 213 cm was used to edge a regular octagon. If the sides of the octagon are the same length as those of the decagon, how much edging was used for the octagon? ______________________

10. A garden 40 meters wide and 72 meters long is to be enclosed by a fence. At $8.90 per meter, what will the fencing cost? ______________________

11. In a class of 32 students, each student needs yarn to edge an art project. About how many yards of yarn are needed for the class if the projects are squares, 8 inches on a side. ______________________

12. David walked around a rectangular field 3 times. How far did he walk if the field's dimensions are 65 meters by 55 meters? ______________________

***Use with Lesson 11-1, text pages 304–305.**

Area: Rectangles, Squares, Parallelograms*

Name ______________________

Date ______________________

Find the area of each.

1. Rectangle
   length: 16 cm
   width: 15 cm

   A = ________

2. Rectangle
   length: 8 cm
   width: 13.4 cm

   A = ________

3. Square
   side: $8\frac{1}{3}$ in.

   A = ________

Complete each chart.

Rectangle

	length	width	Area
4.	14 cm	8.1 cm	
5.	6.3 m	4.2 m	
6.	90 cm	52 cm	
7.	7 ft	$6\frac{1}{3}$ ft	
8.	4 yd	2 ft	

Parallelogram

	base	height	Area
9.	21.3 m	10 m	
10.	2 m	90 cm	
11.	6 dm	1.2 m	
12.	9 in.	3.5 in.	
13.	5 yd	12 ft	

Solve:

14. How many square yards of floor space are there in a cafeteria 23 yd by 18 yd? ________________

15. The janitor is painting the floor of a room 25 m by 40 m. How many containers of paint will he need if one container covers 50 m^2? ________________

16. The neighborhood playground is shaped like a parallelogram with a base of 130 ft and a height of 95 ft. What is the area of the playground? ________________

17. How many square meters of plastic will be needed to make 8 place mats each of which is 40 cm long and 30 cm wide? ________________

18. A square parking lot measures approximately 0.8 km on a side. Estimate the area of the lot. ________________

19. Which has a larger area, a square 12 ft on a side or a rectangle 10 ft long and 14 ft wide? ________________

20. How many tiles 3 in. on a side are needed to cover a space $5\frac{1}{4}$ ft by $3\frac{1}{2}$ ft? ________________

***Use with Lesson 11-2, text pages 306–307.**

Area: Triangles*

Name ____________________

Date ____________________

Find the area of each triangle.

1.

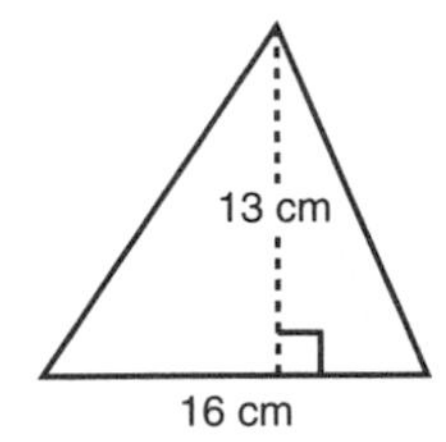

A = ________

2.

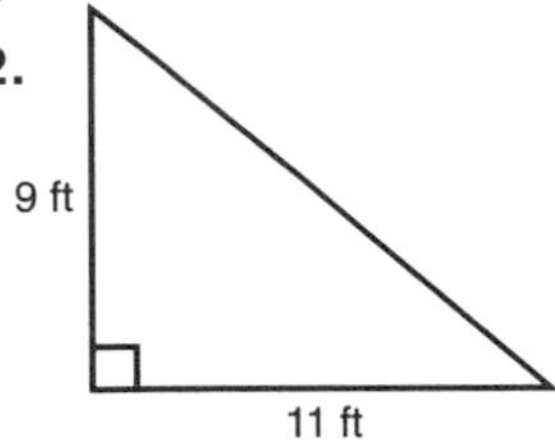

A = ________

3.

19.8 cm

8 cm

A = ________

4.

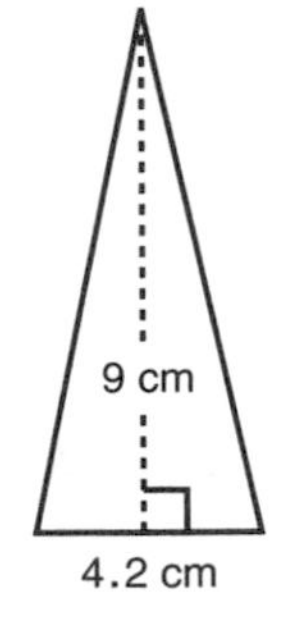

A = ________

Complete the chart for each triangle.

	base	height	Area
5.	4 m	3 m	
6.	18 cm	31 cm	
7.	3 dm	5.6 dm	
8.	40 m	28 m	
9.	52 cm	90 cm	
10.	3 km	1.6 km	
11.	$6\frac{1}{3}$ yd	15 yd	
12.	15 ft	18 ft	

Solve.

13. Find the area of a triangular sail if the base is 3 m and the height is 5 m. ____________________

14. How many square meters of decorative paper would be used for a design made of 6 triangles, each of which has a base of 60 cm and a height of 42 cm? ____________________

15. A gardener planted flowers in a triangular corner of his garden. What is the area of the flower bed if its base is 20 ft and its height is 16.5 ft? ____________________

16. The perimeter of an equilateral triangle is 18 cm and the height is 5.2 cm. What is the area of the triangle? ____________________

***Use with Lesson 11-3, text pages 308–309.**

Area: Trapezoids, Mixed Polygons*

Name ______________________

Date ______________________

Find the area of each.

1.

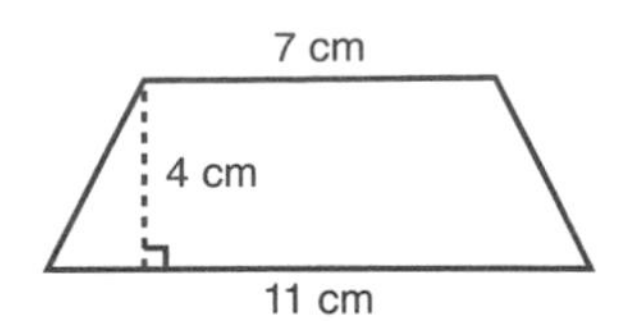

$A =$ ________

2.

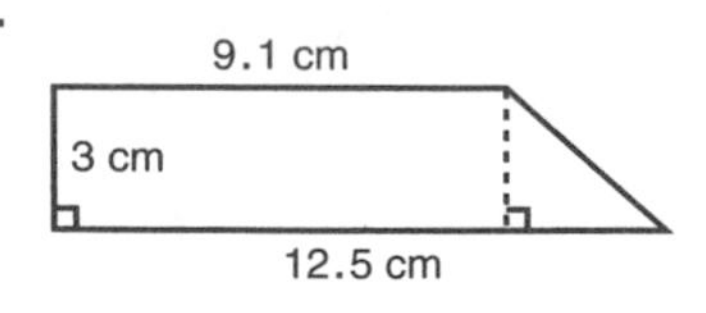

$A =$ ________

3.

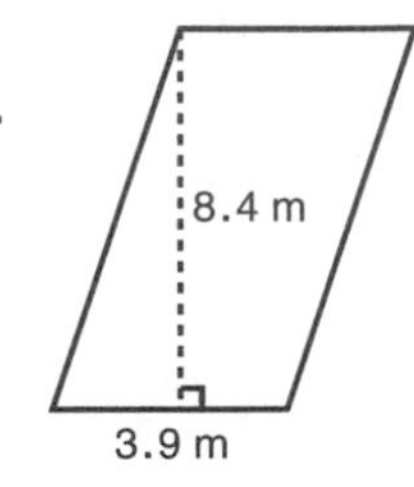

$A =$ ________

Find the area of each trapezoid. Write the area in the unit of measure indicated. Change units where necessary.

4. bases: 9 cm, 23 mm
height: 8 cm

Area: ________ cm^2

5. bases: 4 ft, 2 yd
height: 2 ft

Area: ________ yd^2

6. bases: 4 m, 600 cm
height: 30 m

Area: ________ m^2

Solve:

7. A tabletop shaped like a trapezoid has parallel bases of 40 cm and 62 cm and a height of 34 cm. Find the area. ________

8. Lauren placed a square lamp, measuring 25 cm on each edge, on a table shaped like a trapezoid. The table had parallel bases of 32 cm and 48 cm and a height of 34 cm. Find the number of square centimeters not covered by the lamp. ________

9. A bulletin board is shaped like a trapezoid. If the bases are 45 cm and 65 cm and the height is 40 cm, what is the area? ________

10. Find the area of a trapezoid having a height of 32 in. and parallel bases measuring 49 in. and 61 in. ________

11. A trapezoid has a height of 3 yd and parallel bases of 7 ft and 10 ft. What is its area? ________

12. A line segment drawn parallel to the 12-cm height of a right triangle divides the base into segments of 3 cm and 6 cm and the hypotenuse into segments of 5 cm and 10 cm. This produces a new right triangle and a trapezoid. Find the area of each. (Hint: Use proportion to find the length of the missing side/base. Two sets of answers to the area problem are possible.) ________

***Use with Lesson 11-4, text pages 310–311.**

Circumference*

Name ______________________

Date ______________________

Complete each table.

Use 3.14 for π.

	Diameter	Circumference
1.	3 m	
2.	22 cm	
3.	30 mm	
4.	8.1 m	
5.	1.6 km	
6.	9.3 dm	

Use $\frac{22}{7}$ for π.

	Diameter	Circumference
7.	7 cm	
8.	21.7 cm	
9.	42 dm	
10.	3.5 km	
11.	5.6 m	
12.	$3\frac{1}{2}$ ft	

Find the distance around each figure.

13.

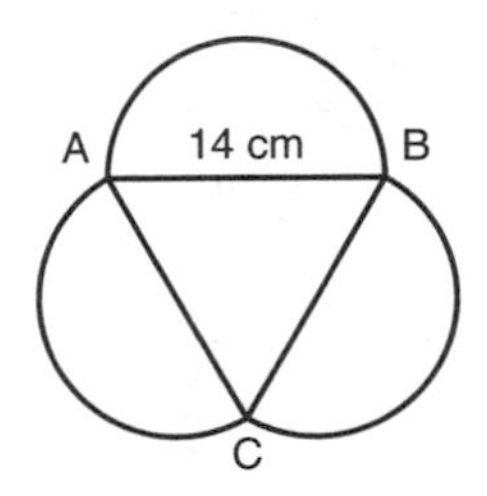

Inside the figure, ABC is an equilateral triangle. Semicircles are positioned on each side.

____

14.

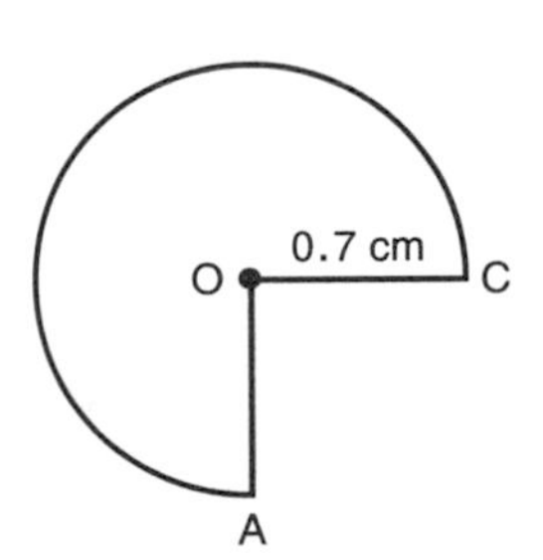

$m\angle AOC = 90°$
$\overline{OC} = 0.7$ cm

____

Solve:

15. How much fencing is needed to enclose a circular garden whose radius is 4.5 yards? ______________________

16. Find the circumference of a bicycle wheel if one spoke from the center measures 24 cm. ______________________

17. A gardener measured 4.2 m as the diameter of a circular flower bed. What was the circumference of the flower bed? ______________________

*Use with Lesson 11-5, text pages 312–313.

Area: Circles*

Name ______________________

Date ______________________

Remember the formula: $A = \pi r^2$

Give the area in the specified square units. Change units where necessary.

1. $r = 7$ cm
$A \approx$ ________ cm^2

2. $r = 4$ cm
$A \approx$ ________ cm^2

3. $r = 12$ m
$A \approx$ ________ m^2

4. $r = 41$ m
$A \approx$ ________ m^2

5. $d = 50$ ft
$A \approx$ ________ yd^2

6. $d = 11$ m
$A \approx$ ________ m^2

7. $d = 40$ mm
$A \approx$ ________ cm^2

8. $d = 2.5$ mm
$A \approx$ ________ mm^2

Find the area.

9.

____

10.

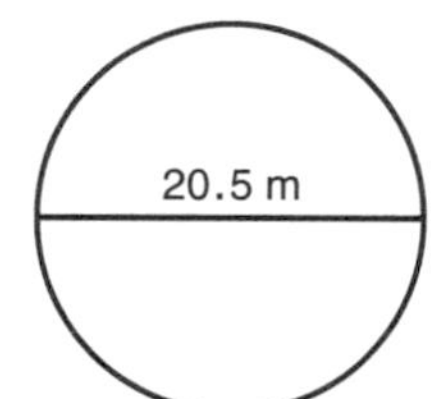

____

11.

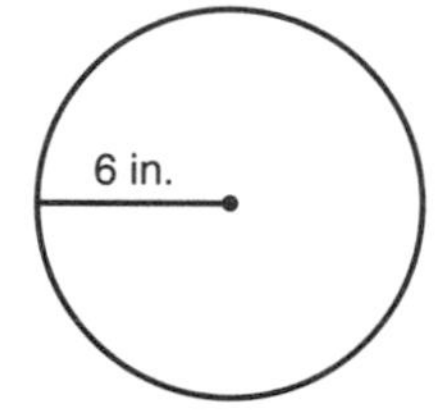

____

12.

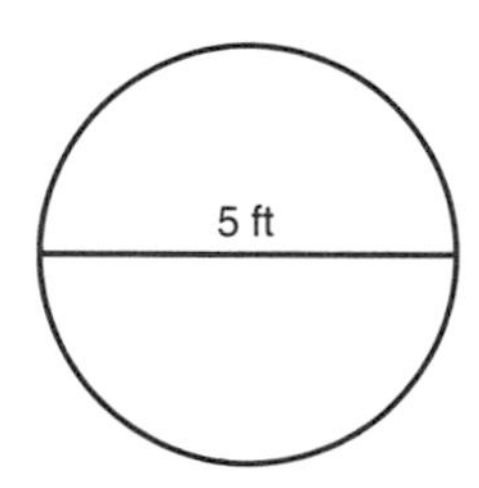

____

13.

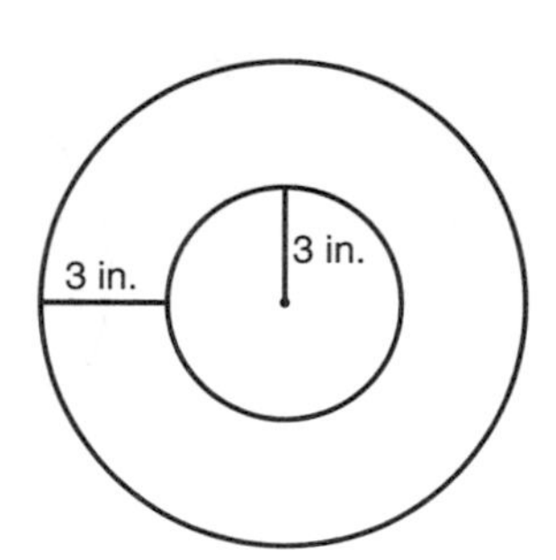

____

14.

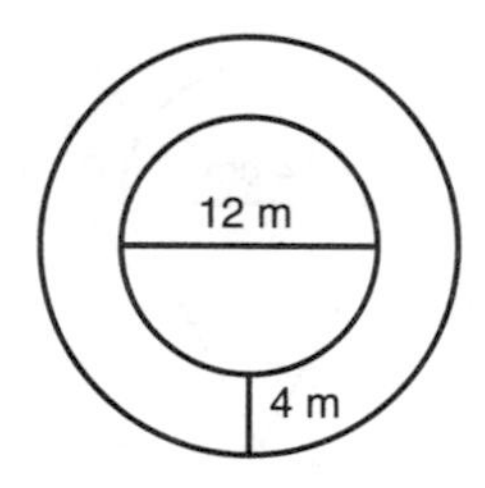

____

Use a calculator to find these areas.

15. $d = 7.02$ m ________

16. $r = 0.015$ m ________

17. $d = 0.529$ cm ________

18. $d = 4.12$ m ________

19. $d = 426$ km ________

20. $r = 42.8$ m ________

21. $d = 96$ m ________

22. $d = 6.054$ m ________

23. How many times larger is the area of a circle with a diameter of 20 cm than one with a diameter of 5 cm? ________

24. The radius of an ice-skating rink is 75 meters. How many square meters of ice surface will it take to cover it? ________

25. A semicircle has a diameter of 8.7 m. What is its area? ________

*Use with Lesson 11-6, text pages 314–315.

Area: Irregular Figures*

Name ______________________

Date ______________________

Find the area of each figure.

1.

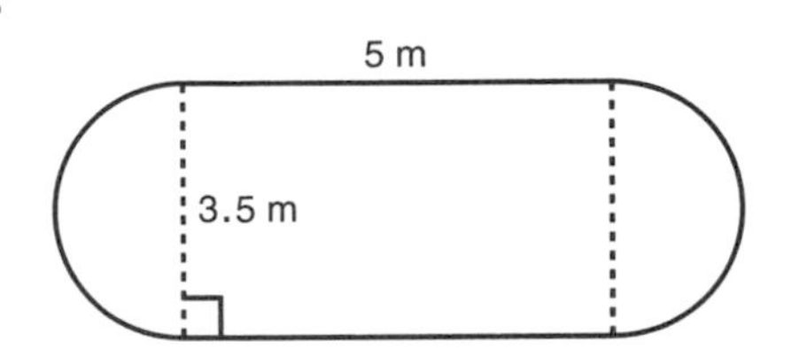

$A \approx$ __________

2.

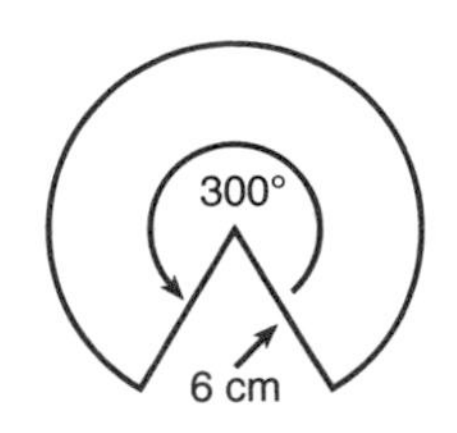

$A \approx$ __________

3.

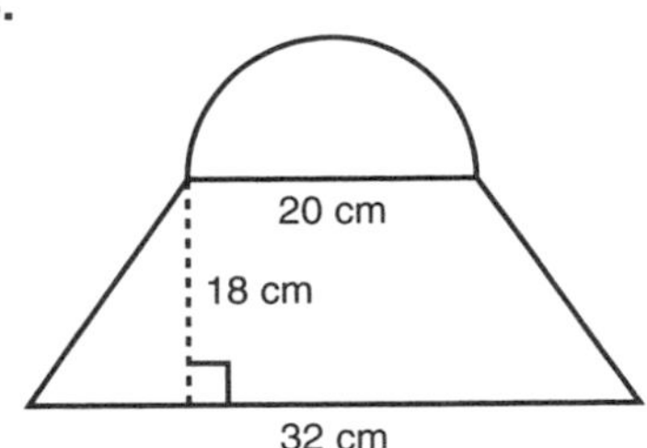

$A \approx$ __________

Find the area of the *shaded region* of each figure.

4.

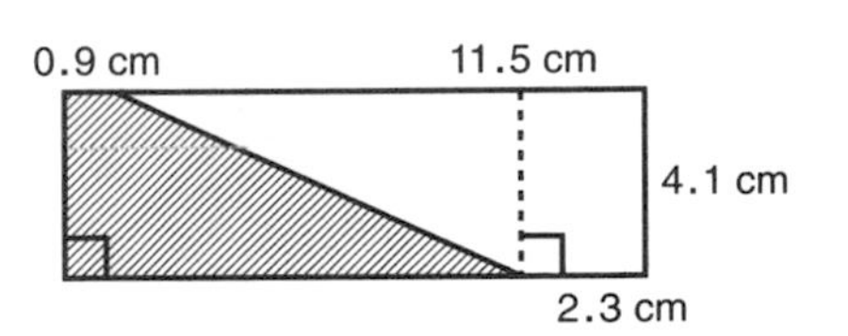

$A =$ __________

5.

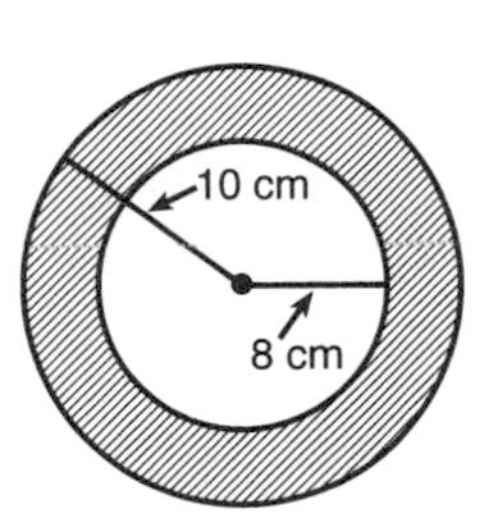

$A \approx$ __________

6.

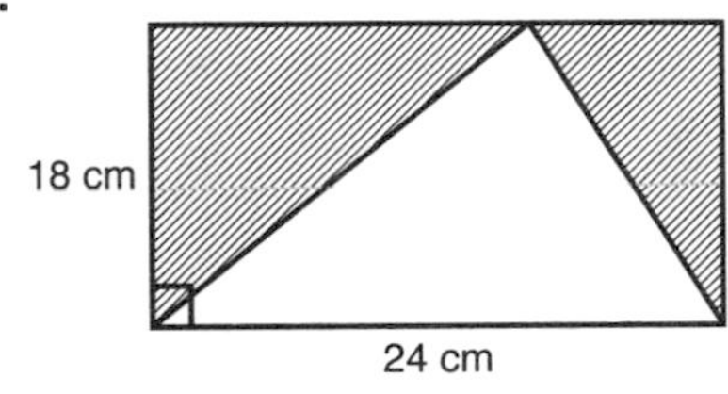

$A =$ __________

7.

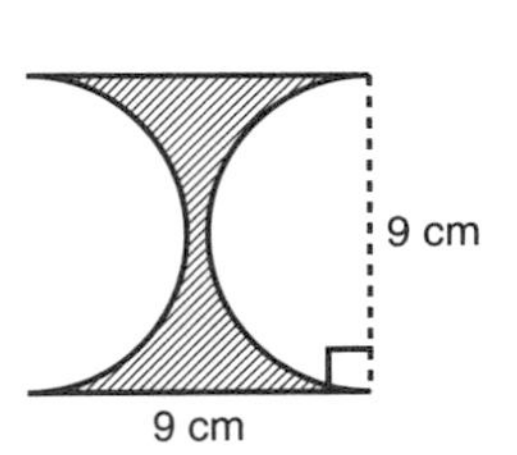

$A \approx$ __________

8.

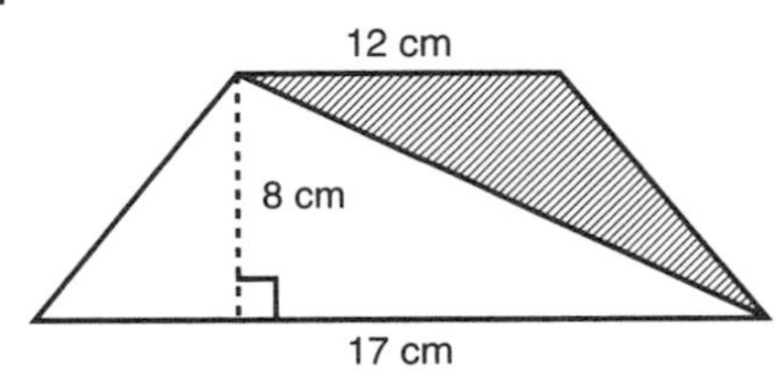

$A =$ __________

9.

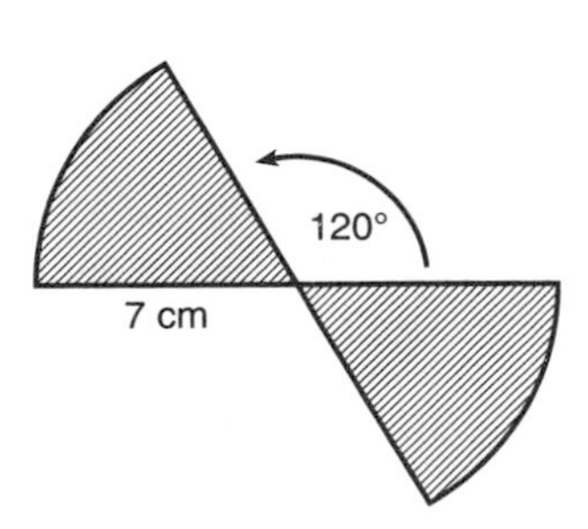

$A \approx$ __________

*Use with Lesson 11-7, text pages 316–317.

Missing Dimensions*

Name ____________

Date ____________

Find the missing dimension.

1. Rectangle: $A = 229.2 \text{ m}^2$
   length: 19.1 m
   width: ______

2. Triangle: $A = 1024 \text{ yd}^2$
   height: 20 yd
   base: ______

3. Square: $A = 2.25 \text{ ft}^2$
   side: ______

Complete each of the following charts.

Rectangle

	length	width	Area
4.	78 cm	53 cm	
5.		0.72 m	0.648 m^2
6.	$4\frac{1}{2}$ ft		3 ft^2
7.		$1\frac{1}{3}$ ft	1 yd^2
8.	500 m		1 ha

Parallelogram

	length	width	Area
9.	9.1 cm	8 cm	
10.	78 cm		4914 cm^2
11.		$6\frac{1}{3}$ yd	$28\frac{1}{2} \text{ yd}^2$
12.		12 ft	102 ft^2
13.	500 m		15 ha

Square

	side	Area
14.	1.3 m	
15.	18 ft	
16.		2.89 ft^2
17.		196 yd^2
18.		2500m^2

Triangle

	base	height	Area
19.	13 cm	18.6 cm	
20.		5.2 m	62.4 m^2
21.	$8\frac{1}{3}$ ft		$8\frac{1}{3} \text{ ft}^2$
22.		24 ft	16 yd^2
23.	2.4 yd		3.6 yd^2

Solve.

24. One rectangular section of a farm covers an area of 3 hectares. If the length of this section is 250 m, what is its width? ______

25. The area of a triangular stained glass window is 0.315 m^2. The base of the window measures 0.7 m. What is its height? ______

26. A lamp with a circular base 14 cm in diameter was on a square mat with an area of 196 cm^2. How many cm^2 of the mat were uncovered? ______

*Use with Lesson 11-9, text pages 320–321.

Repeating Decimals As Fractions*

Name ______________________

Date ______________________

Change each to the fractional form.

1. $71 =$ ________ **2.** $^{-}0.9 =$ ________ **3.** $2\frac{4}{5} =$ ________ **4.** $^{-}1.7 =$ ________

5. $^{-}314 =$ ________ **6.** $0.08 =$ ________ **7.** $3.06 =$ ________ **8.** $0.003 =$ ________

Complete.

9.

$x = 0.\overline{63}$

$100x = 0.\overline{63} \times$ ________

$100x =$ ________

$-x =$ ________

$99x =$ ________

$x =$ ________

10.

$x = 0.08\overline{3}$

$100x = 8.\overline{3}$

____$x = 83.\overline{3}$

$-$____$x = 8.\overline{3}$

$900x =$ ________

$x =$ ________

Change each to a fraction.

11. $0.\overline{16} =$ ________ **12.** $0.\overline{09} =$ ________ **13.** $0.\overline{1} =$ ________ **14.** $0.\overline{25} =$ ________

15. $0.8\overline{7} =$ ________ **16.** $0.1\overline{6} =$ ________ **17.** $0.8\overline{3} =$ ________ **18.** $0.\overline{81} =$ ________

19. $0.\overline{7} =$ ________ **20.** $0.\overline{18} =$ ________ **21.** $0.41\overline{6} =$ ________ **22.** $0.\overline{26} =$ ________

*Use with Lesson 12-1, text pages 332–333.

Squares, Square Roots, Irrational Numbers*

Name ______________________

Date ______________________

Find the square.

1. 0.5 ________
2. 0.7 ________
3. 0.06 ________
4. $\frac{1}{4}$ ________
5. $\frac{2}{7}$ ________
6. $\frac{4}{5}$ ________
7. 0.9 ________
8. 0.04 ________
9. $\frac{6}{7}$ ________
10. $\frac{3}{4}$ ________
11. 0.8 ________
12. 0.02 ________
13. 0.09 ________
14. $\frac{5}{6}$ ________
15. $\frac{2}{9}$ ________
16. $\frac{5}{8}$ ________
17. 0.07 ________
18. $\frac{1}{8}$ ________
19. 0.05 ________
20. $\frac{3}{5}$ ________

This table shows the squares of whole numbers from 1–20. Use it to find each square root below.

Number	1	2	3	4	5	6	7	8	9	10	11	12	13	14	15	16	17	18	19	20
Square	1	4	9	16	25	36	49	64	81	100	121	144	169	196	225	256	289	324	361	400

21. $\sqrt{\frac{49}{144}}$ = ________
22. $\sqrt{\frac{16}{225}}$ = ________
23. $\sqrt{\frac{81}{256}}$ = ________
24. $\sqrt{\frac{169}{400}}$ = ________
25. $\sqrt{3.61}$ = ________
26. $\sqrt{0.36}$ = ________
27. $\sqrt{2.89}$ = ________
28. $\sqrt{1.69}$ = ________
29. $\sqrt{2.25}$ = ________
30. $\sqrt{\frac{49}{324}}$ = ________
31. $\sqrt{\frac{25}{121}}$ = ________
32. $\sqrt{1.96}$ = ________

Between which two whole numbers would the square root be?

33. $\sqrt{7}$ ________
34. $\sqrt{12}$ ________
35. $\sqrt{23}$ ________
36. $\sqrt{35}$ ________
37. $\sqrt{142}$ ________
38. $\sqrt{66}$ ________
39. $\sqrt{300}$ ________
40. $\sqrt{72}$ ________
41. $\sqrt{120}$ ________
42. $\sqrt{250}$ ________
43. $\sqrt{350}$ ________
44. $\sqrt{5}$ ________
45. $\sqrt{69}$ ________
46. $\sqrt{299}$ ________
47. $\sqrt{170}$ ________
48. $\sqrt{150}$ ________
49. $\sqrt{168}$ ________
50. $\sqrt{380.6}$ ________

*Use with Lesson 12-2, text pages 334–335.

Square-Root Algorithm*

Name ________________

Date ________________

Complete to find the square root.

1.

```
       6'05'16.|2 _ _
  2  | -4
       2 05
  4_ | -1 76
         29 16
_ _ _|  -29 16
```

2.

```
       21'90'24.|4 _ _
  4  | -16
        5 90
  8_ |  -5 16
         74 24
_ _ _|  -74 24
```

3.

```
       6'05.16'|_ _ _
  2  | -4
       2 05
  4_ | -1 76
         29 16
_ _ _|  -29 16
```

Find the square root of each of the following.

4. $\sqrt{3025}$ **5.** $\sqrt{2304}$ **6.** $\sqrt{8281}$ **7.** $\sqrt{6889}$

8. $\sqrt{11{,}664}$ **9.** $\sqrt{55{,}225}$ **10.** $\sqrt{66{,}049}$ **11.** $\sqrt{90{,}601}$

12. $\sqrt{0.0784}$ **13.** $\sqrt{36.4816}$ **14.** $\sqrt{800.89}$ **15.** $\sqrt{0.4761}$

16. $\sqrt{9350.89}$ **17.** $\sqrt{1989.16}$ **18.** $\sqrt{942.49}$ **19.** $\sqrt{1310.44}$

***Use with Lesson 12-3, text pages 336–337.**

Real Number System*

Name ______________________

Date ______________________

Remember: Every point on a number line can be associated with a real number.

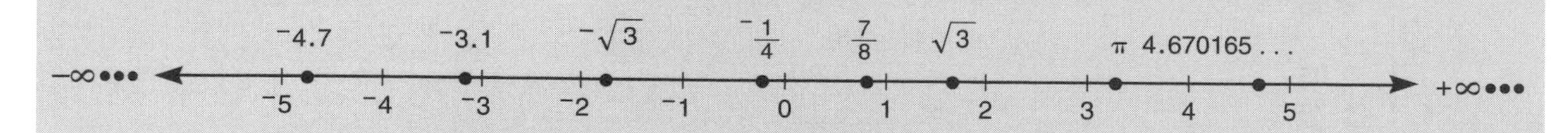

Classify each real number as rational or irrational.

1. 6 ______________________

2. 0.4321 ______________________

3. $\sqrt{12}$ ______________________

4. $(^+6.3)^2$ ______________________

5. 0.2 ______________________

6. $^-0.7\overline{07}$ ______________________

7. $^-17$ ______________________

8. $^-\sqrt{5}$ ______________________

9. $2\frac{4}{5}$ ______________________

10. $(^-0.3)^2$ ______________________

Solve.

11. If $a^2 = 16$, then $a =$ ______________________

12. If $b^2 = 121$, then $b =$ ______________________

13. If $y^2 = 64$, then $y =$ ______________________

14. If $x^2 = 361$, then $x =$ ______________________

15. If $c^2 = 49$, then $c =$ ______________________

16. If $d^2 = 196$, then $d =$ ______________________

Find the midpoint between each pair of numbers.

17. 0.2 and 0.3 ______________________

18. 3.6 and 3.71 ______________________

19. $\frac{5}{7}$ and $\frac{6}{7}$ ______________________

20. $8\frac{1}{2}$ and $8\frac{2}{3}$ ______________________

21. 0.008 and 0.009 ______________________

22. $^-0.36$ and $^-0.37$ ______________________

Name two irrational numbers between each pair.

23. 7 and 8 ______________________

24. 12 and 13 ______________________

25. 5 and 6 ______________________

26. 15 and 16 ______________________

27. 17 and 18 ______________________

28. 19 and 20 ______________________

*Use with Lesson 12-4, text pages 338–339.

Real Number Solutions for Inequalities*

Name ______________________

Date ______________________

$r + 7 > 10.5$

$r + 7 - 7 > 10.5 - 7$

$r > 3.5 \rightarrow \{r : r > 3.5\}$

Write the solution set.

1. $r - {}^{-}11 > 3$ ______________________

2. $7n + {}^{-}2 > {}^{-}16$ ______________________

3. $11x - 5 \leq 50$ ______________________

4. $\frac{a}{8} + 1 \leq 2$ ______________________

5. $\frac{b}{3} - {}^{-}3 \geq 5$ ______________________

6. $4a - 7 > 21$ ______________________

7. $\frac{x}{4} + {}^{-}5 \leq 7$ ______________________

8. $6m + 11 \geq 41$ ______________________

9. $10t + 9 < 69$ ______________________

10. $\frac{b}{12} - 3 \leq 2$ ______________________

11. $\frac{y}{3} \leq {}^{-}19$ ______________________

12. $33x \neq 2508$ ______________________

13. $12a < 252$ ______________________

14. $q + 9 \leq {}^{-}18$ ______________________

15. $k - 4.8 \leq 6.3$ ______________________

16. $1.2b > 4.8$ ______________________

17. $a - 7 \neq {}^{-}3$ ______________________

18. $\frac{y}{3} \leq 2$ ______________________

19. $\frac{2r}{3} \leq \frac{4}{5}$ ______________________

20. $\frac{b}{2.7} \geq 0.3$ ______________________

***Use with Lesson 12-5, text pages 340–341.**

The Pythagorean Theorem*

Name ______________________

Date ______________________

Find the missing dimension for each right triangle.

1.

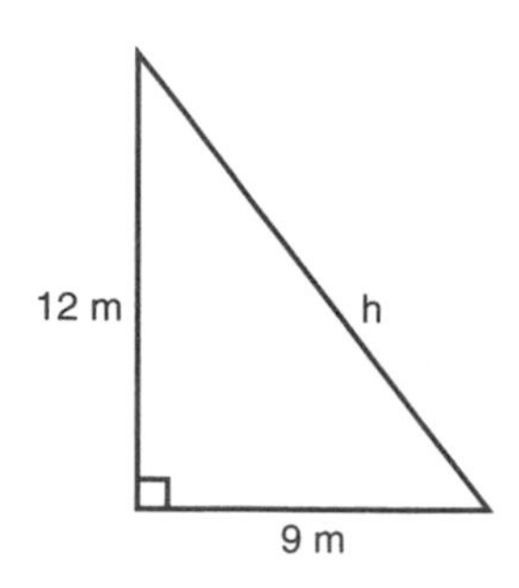

h = ____________

2.

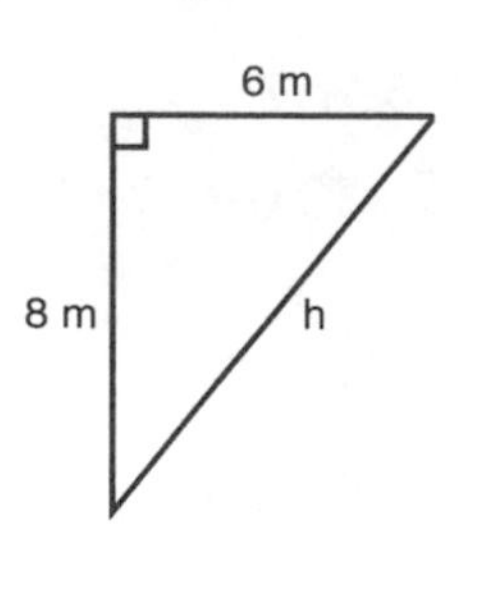

h = ____________

3.

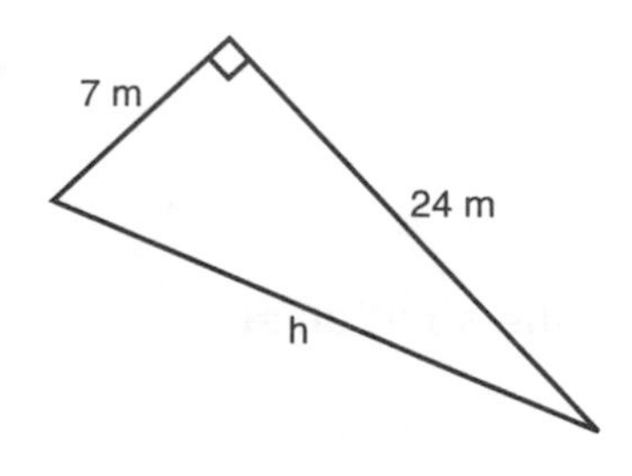

h = ____________

4.

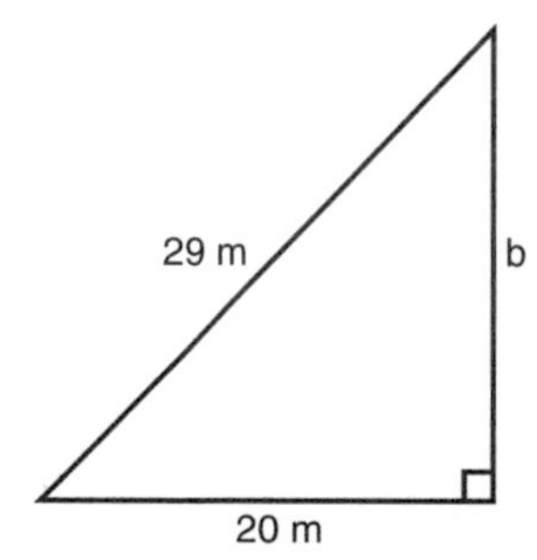

b = ____________

5.

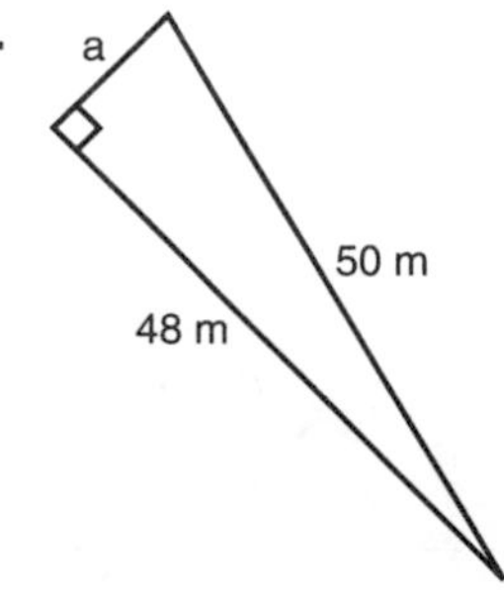

a = ____________

6.

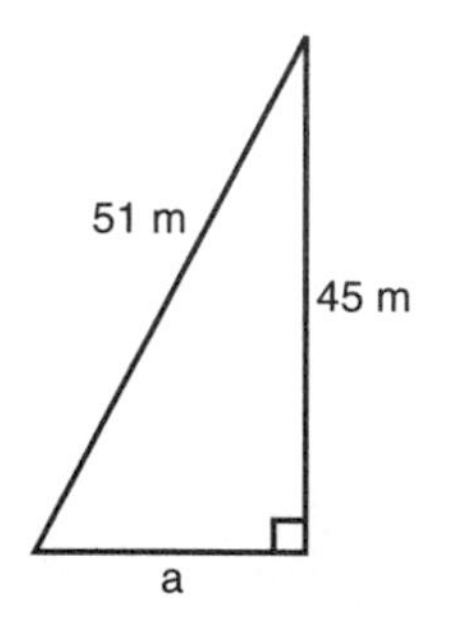

a = ____________

Complete. Find the missing dimension for each right triangle.

	7.	8.	9.	10.
hypotenuse	105 m		250 m	130 m
side a	63 m	120 cm	150 m	
side b		160 cm		104 m

Draw a picture and solve.

11. A rectangular lot measures 12 m by 16 m. How long is the walk that runs diagonally across it? ____________

12. A 17-ft ladder is leaning against a house. The top of the ladder touches the house at a point 15 ft above the ground. How far out from the house is the foot of the ladder. ____________

***Use with Lesson 12-6, text pages 342–343.**

Surface Area: Rectangular Prisms and Cubes*

Name ______________________

Date ______________________

Find the surface area of each rectangular prism. $S = 2[(\ell w + (\ell h) + (wh)]$

	length	width	height	Surface Area
1.	10 in.	8 in.	4 in.	
2.	14 ft	9 ft	4 ft	
3.	4 m	4 m	5.2 m	
4.	8 ft	1.6 ft	6 ft	
5.	2.4 cm	4 cm	1.2 cm	
6.	24 ft	14 ft	8 ft	
7.	12 cm	20 cm	6 cm	
8.	2.4 m	1.6 m	4 m	

Find the surface area of a cube with an edge of: $S = 6e^2$

9. 18 in. S = ________

10. 9 ft S = ________

11. 16 ft S = ________

12. 2.4 cm S = ________

Solve.

13. Find the surface area of an aluminum baking pan 25 cm wide, 32 cm long, and 10 cm deep. ________________

14. Estimate to the nearest tenth the number of square yards of fabric needed to line a suitcase 2 ft long, 1 ft 6 in. wide, and 6 in. deep. ________________

15. Find the surface area of a cubical planter measuring 30 in. on a side. (The planter has no lid.) ________________

16. How many square yards of cedar wood will be needed to line the walls and ceiling of a closet 2 yd long, $1\frac{2}{3}$ yd wide, and $2\frac{1}{2}$ yd high? ________________

17. What is the surface area of a utility cabinet 60 cm long, 46 cm wide, and 32 cm high? ________________

18. A cubical ice-cream freezer is 20 cm on a side. What is its surface area? ________________

*Use with Lesson 13-1, text pages 354–355.

Surface Area: Triangular Prisms*

Name ______________________

Date ______________________

Find the surface area of each triangular prism.

$$S = 2\left(\frac{1}{2}bh\right) + (\ell_1 w) + (\ell_2 w) + (\ell_3 w)$$

1.

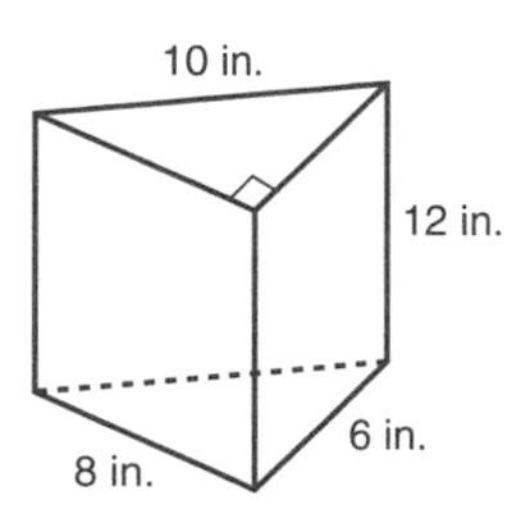

$S =$ __________

2.

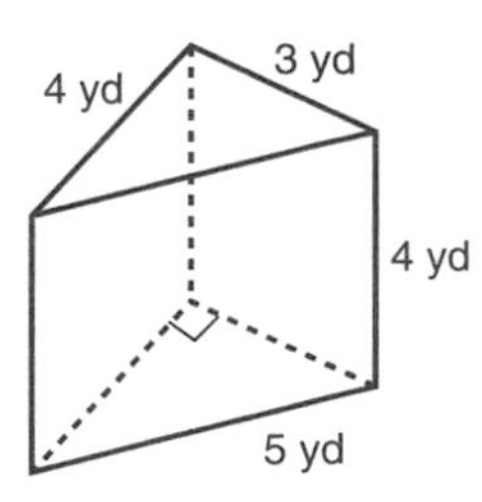

$S =$ __________

3.

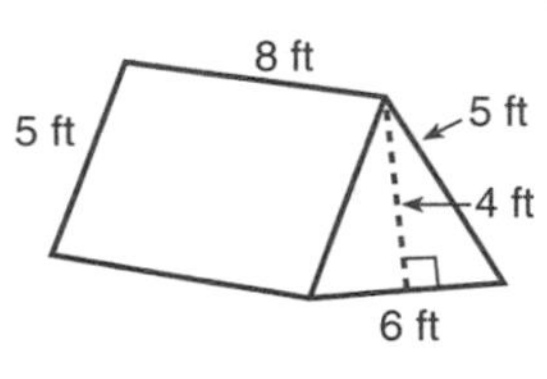

$S =$ __________

4.

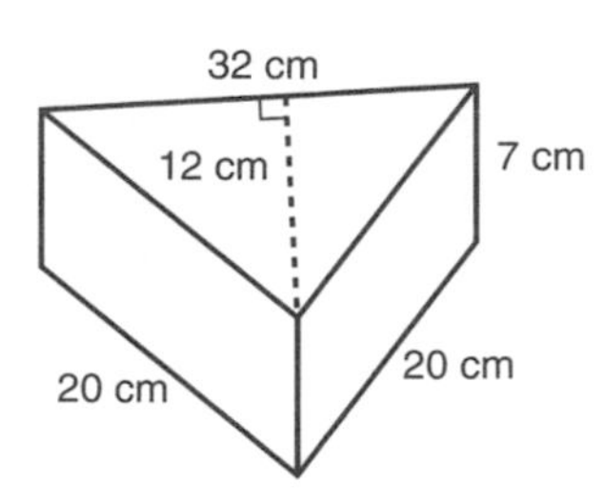

$S =$ __________

5.

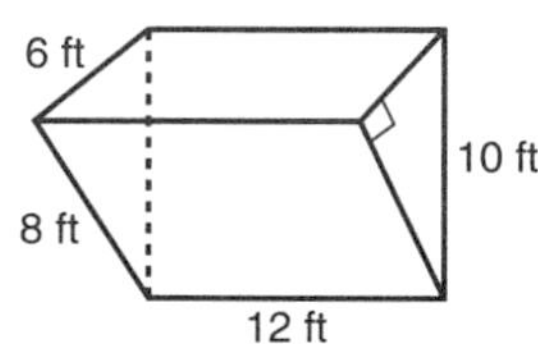

$S =$ __________

6.

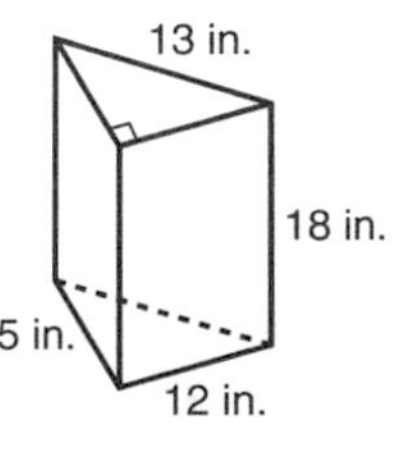

$S =$ __________

7.

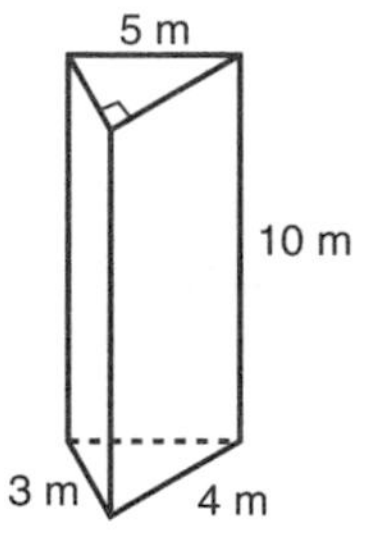

$S =$ __________

8.

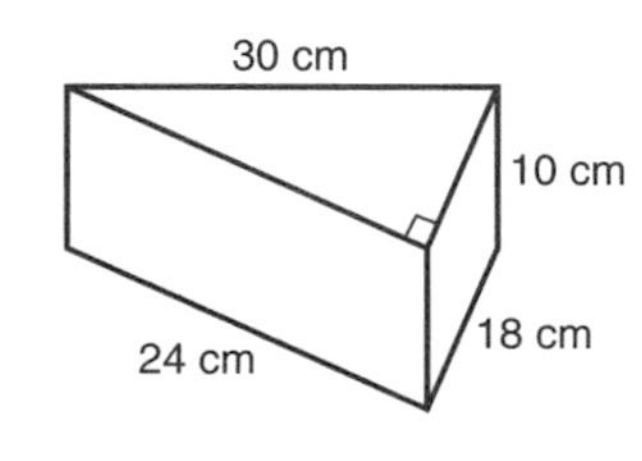

$S =$ __________

Solve.

9. Mr. Grant made a triangular playhouse for his children in one corner of the family room. The triangular base has a height of 1.6 m and a base of 2.1 m. How many square meters of plywood did Mr. Grant use if the playhouse is 1.8 m tall and 2.6 m wide?

____

10. For a geometry project, the math class made space figures. Three students made right triangular prisms, with the dimensions given here. Who used the most paper?

	Stephanie	Yolanda	Joanna
base	16 cm	20 cm	7 cm
height	12 cm	15 cm	24 cm
length	10 cm	22 cm	28 cm
width	20 cm	25 cm	25 cm

____

____

____

*Use with Lesson 13-1, text pages 354–355.

Surface Area: Pyramids*

Name ____________________

Date ____________________

Find the surface area of each pyramid.

S = The sum of the areas of all the surfaces

1.

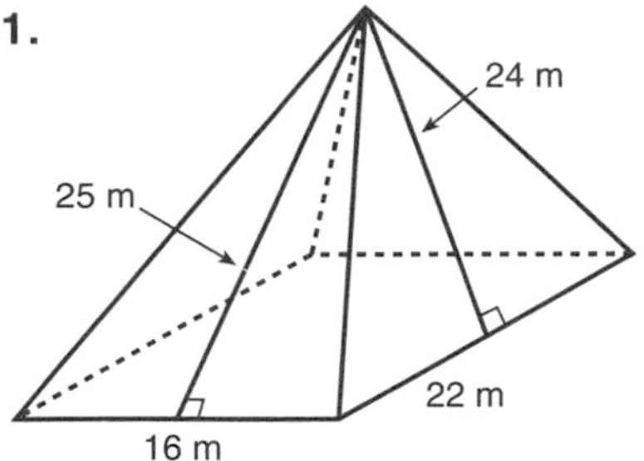

S = ________

2.

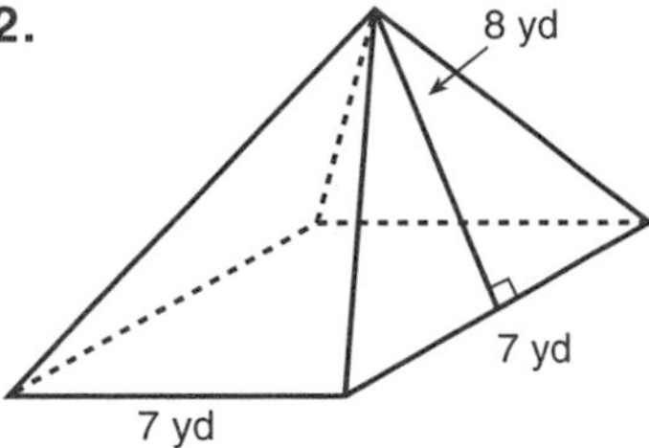

S = ________

3. The faces are congruent equilateral triangles.

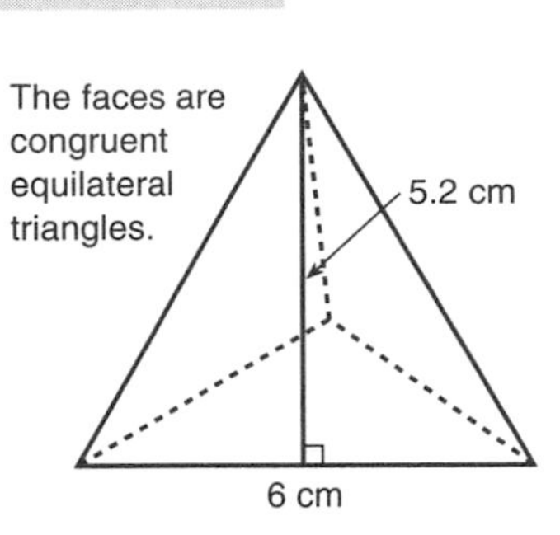

S = ________

4. The base is a regular hexagon with an area of $10.4\ cm^2$.

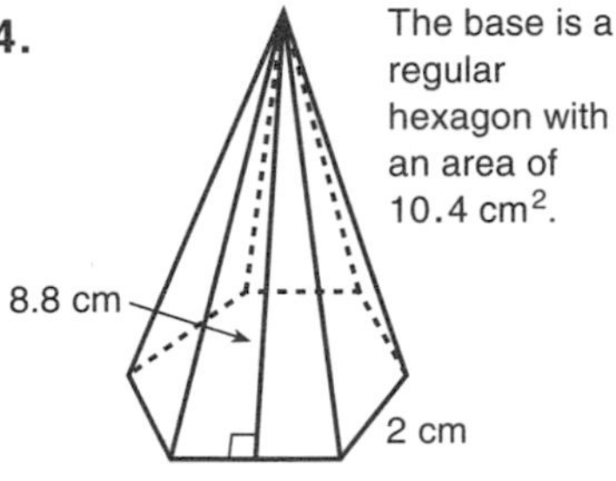

S = ________

5.

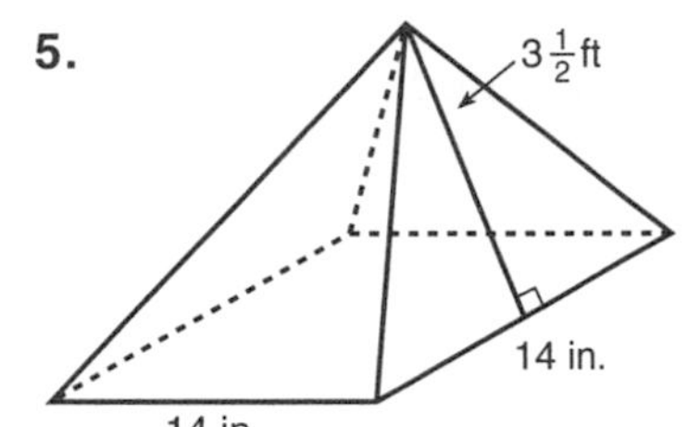

S = ________

6. The base is an equilateral triangle.

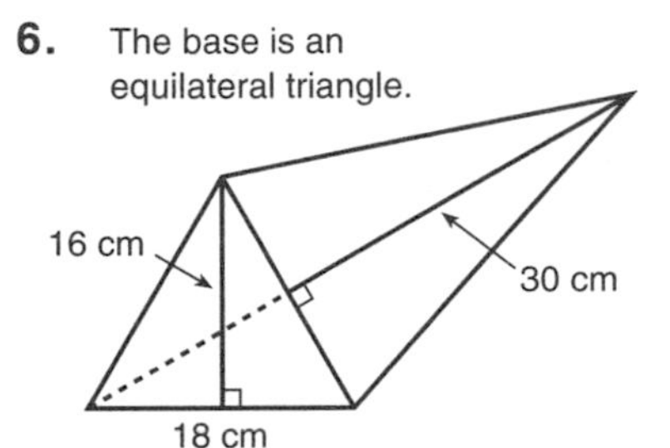

S = ________

Solve.

7. How many square centimeters of colored paper did Jane use in constructing a rectangular pyramid with base measurements of 20 cm by 16 cm and triangle heights of 22 cm? ____________________

8. Ryan made a clear plastic case shaped like a square pyramid to enclose a prized athletic trophy. Find the surface area of the case if the edge of the base measures 30 cm and the height of each triangle is 25 cm. ____________________

9. Dolores mistakenly ordered sturdy plastic material to make 3 cubes 20 cm on a side. She should have ordered material for 3 square pyramids formed by a square base 20 cm on a side and triangles having heights of 20 cm. How much plastic did Dolores overorder? ____________________

10. Find the surface area of a triangular pyramid if the area of the base is $6.9\ m^2$, the height of each triangle is 5 m, and each base is 4 m. ____________________

*Use with Lesson 13-2, text pages 356–357.

Surface Area: Cylinders*

Name ______________________

Date ______________________

Find the surface area. Use 3.14 or $\frac{22}{7}$ for π.

$S = (2 \times \pi r^2) + (2\pi r \times h)$

1.

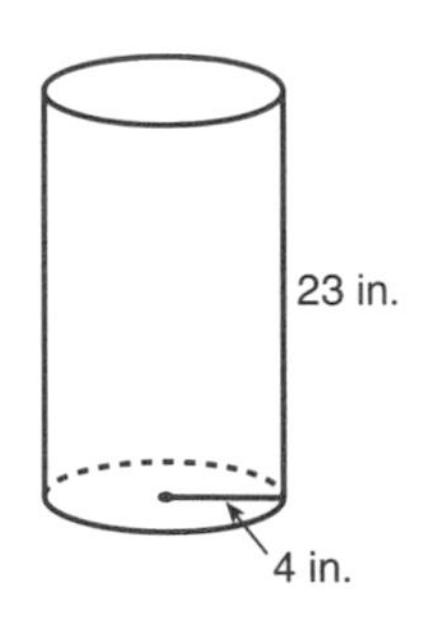

$S \approx$ ________

2.

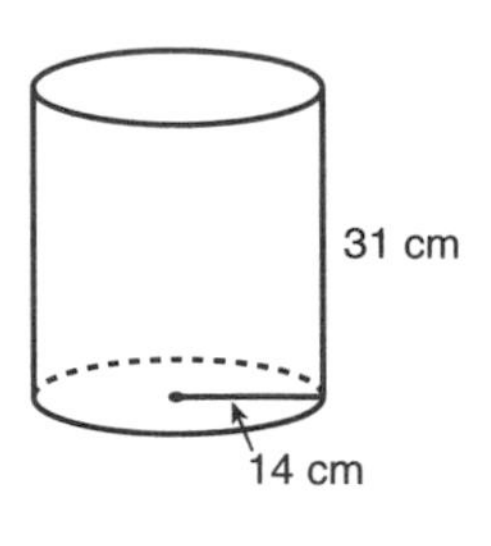

$S \approx$ ________

3.

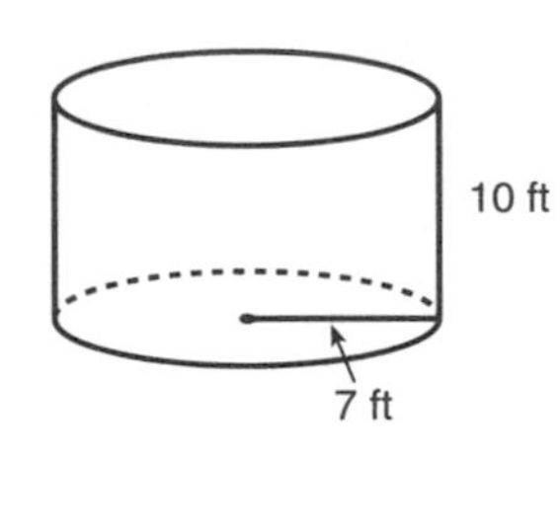

$S \approx$ ________

4.

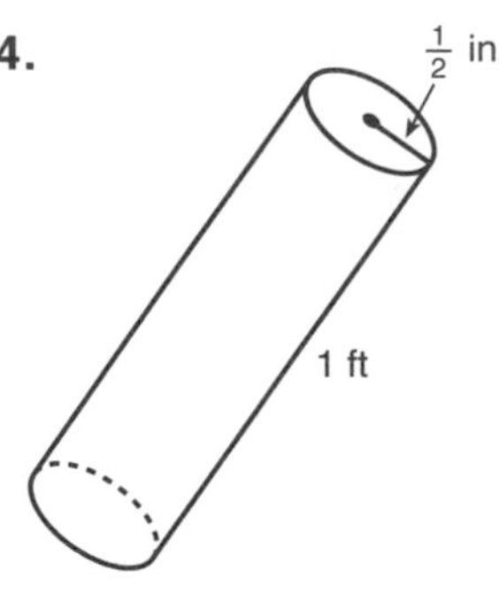

$S \approx$ ________

Complete the table. Use 3.14 or $\frac{22}{7}$ for π as indicated.

	radius	height	π	Surface Area
5.	3 yd	10 yd	3.14	
6.	4 ft	9 ft	3.14	
7.	7 ft	20 ft	$\frac{22}{7}$	
8.	6 m	6 m	3.14	
9.	28 cm	50 cm	$\frac{22}{7}$	
10.	40 cm	10.3 cm	3.14	
11.	5 ft	20 ft	3.14	
12.	$3\frac{1}{2}$ ft	4 ft	$\frac{22}{7}$	
13.	4.2 m	2.6 m	3.14	
14.	9 in.	7 ft	3.14	

Solve.

15. Cans made by a local canning company are 6.4 cm in diameter and 12.5 cm high. How much aluminum is needed to make 100 cans? ______________________

16. A basket is shaped like a cylinder. The diameter is 28 cm and the height is 40 cm. Find the surface area to the nearest tenth square meter. ______________________

*Use with Lesson 13-3, text pages 358–359.

Surface Area: Cones and Spheres*

Name ______________________

Date ______________________

Find the surface area of each cone. $S = \pi r^2 + \pi r \ell$

1.

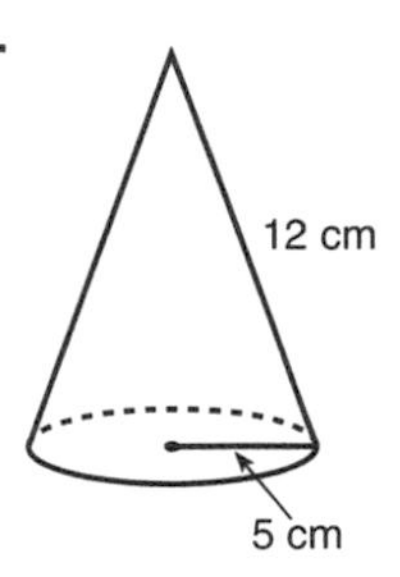

$S \approx$ __________

2.

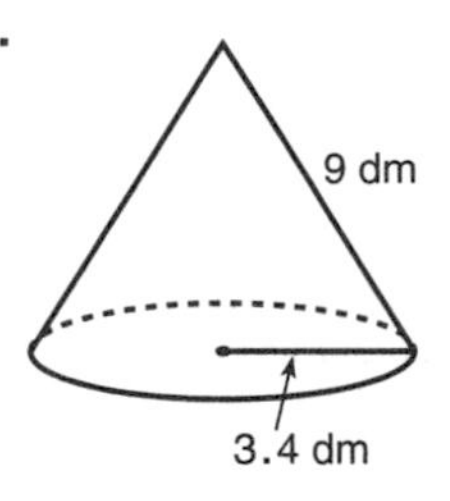

$S \approx$ __________

3.

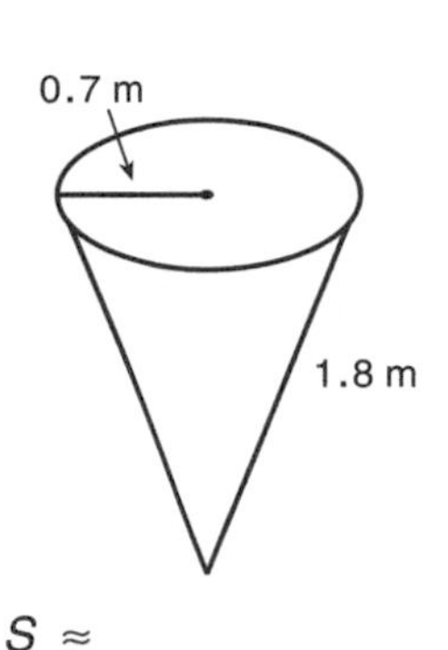

$S \approx$ __________

4.

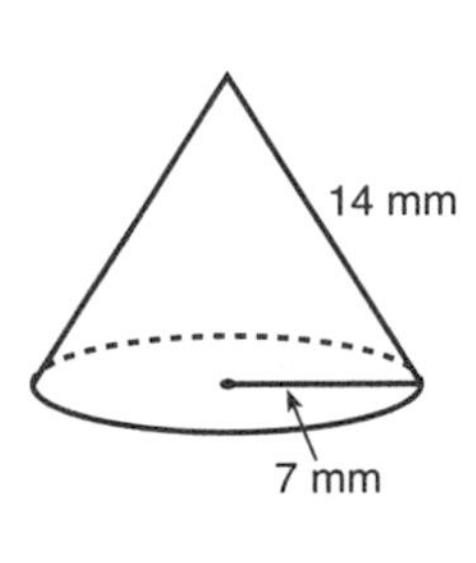

$S \approx$ __________

Find the surface area of each sphere. $S = 4\pi r^2$

5.

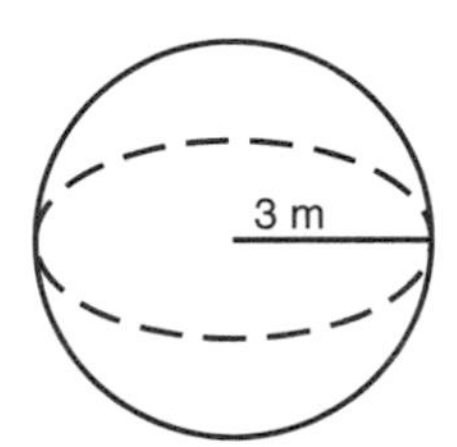

$S \approx$ __________

6.

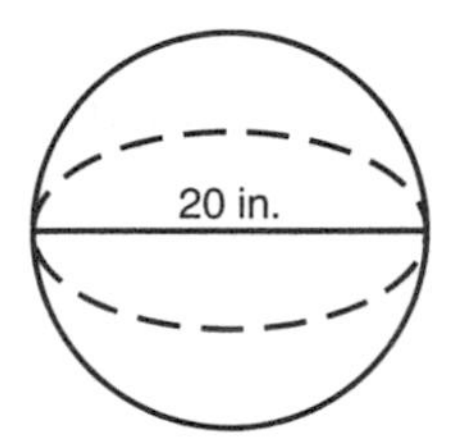

$S \approx$ __________

7.

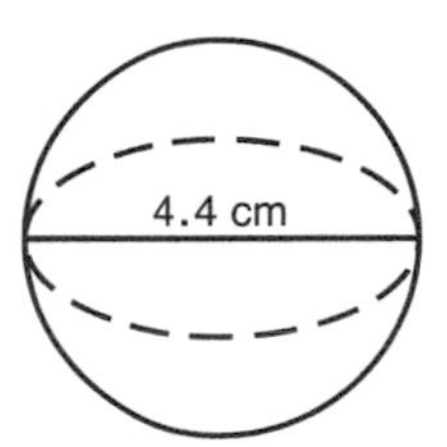

$S \approx$ __________

8.

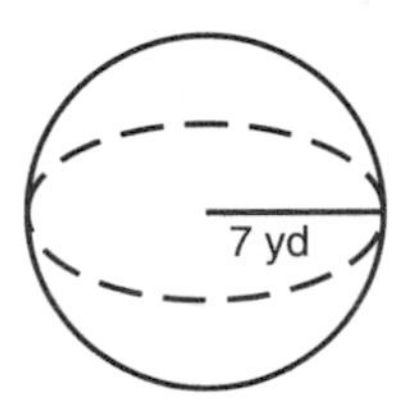

$S \approx$ __________

Solve.

9. How much cardboard is needed to make 6 megaphones each of which has a radius of 32 cm and a slant height of 36 cm? (Find the area of the curved surfaces only.) ______________________

10. Find the surface area of a kickball having an 8-in. diameter. ______________________

11. The entertainment committee made 20 party hats in the shape of cones. Each had a radius of 7 cm and a slant height of 15 cm. Find the total area of the curved surfaces. ______________________

12. Find the surface area of a balloon which, when blown up, has a diameter of 30 cm. ______________________

13. The globe used in social studies class has a diameter of 42 cm. What is its surface area? ______________________

14. How much felt was used to make 10 cone-shaped hats each of which has a diameter of 22 cm and slant height of 30 cm? ______________________

*Use with Lesson 13-3, text pages 358–359.

Volume: Cubes*

Name ____________________

Date ____________________

Find the volume of each cube. $V = e^3$

1.

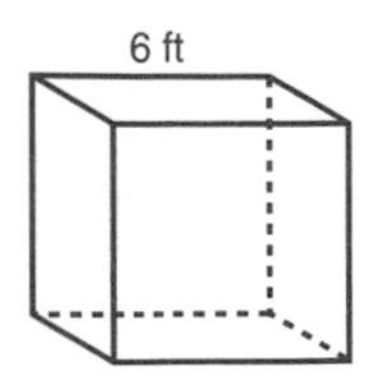

$V =$ ________

2.

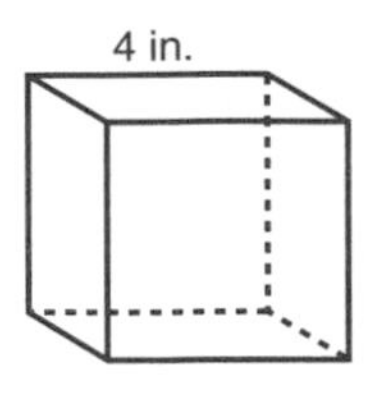

$V =$ ________

3.

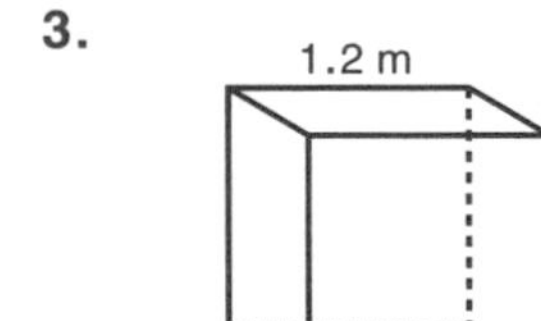

$V =$ ________

4. $1\frac{1}{2}$ ft

$V =$ ________

Examine the cube below and then answer questions 5–9.

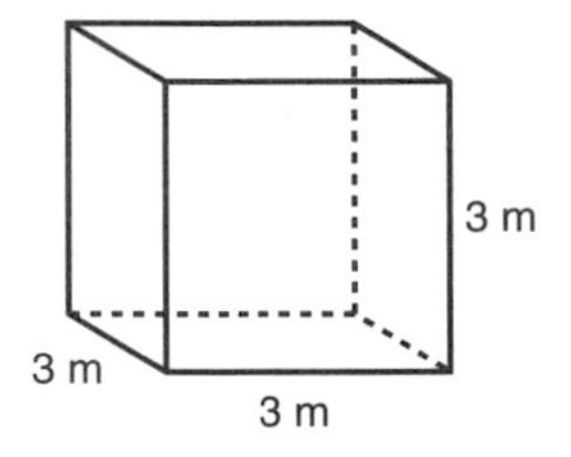

5. What is the total surface area of the cube? ________

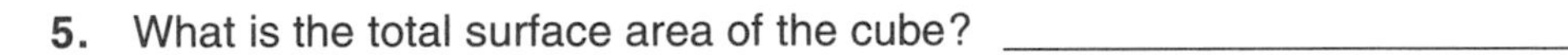

6. What is the volume of the cube? ________

7. If the length of each edge is doubled, what happens to the volume?

____

8. How many cubes 1.5 m on an edge would fit inside this cube, which is 3 m on each edge?

____

9. Could the contents of this cube fit perfectly into a rectangular box? Would the dimensions of the box be unique?

____

Solve.

10. How many cubic feet are there in a block of marble $2\frac{1}{2}$ feet on each edge? ________

11. How many cubes 4 cm on an edge can fit into a cubical box 72 cm on an edge? ________

12. What is the volume of a swimming pool 11 feet wide, 11 feet long, and 11 feet deep? ________

13. A statue is resting on a marble cube 60 cm on an edge. What is the volume of the cube to the nearest cubic meter? ________

14. Find the surface area and the volume of a cubical container 6 ft on an edge. ________

*Use with Lesson 13-4, text pages 360–361.

Volume: Rectangular Prisms*

Name ____________________

Date ____________________

Find the volume of each rectangular prism. $V = Bh$

1.

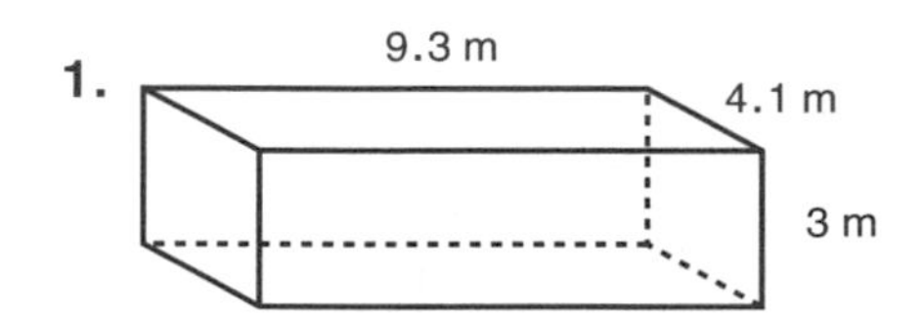

$V =$ __________

2.

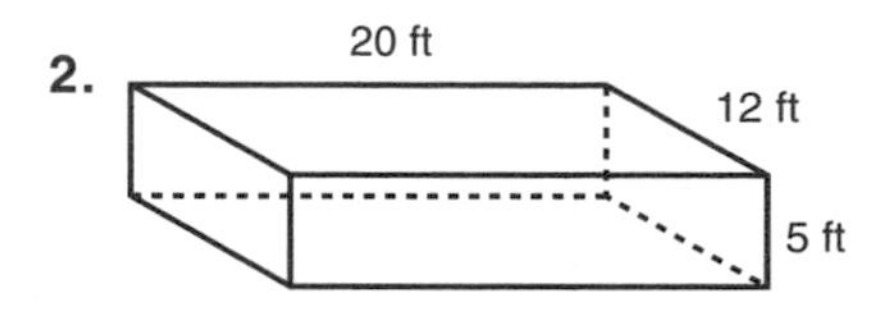

$V =$ __________

3.

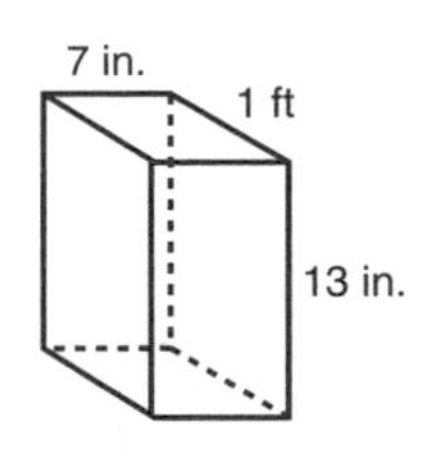

$V =$ __________

4.

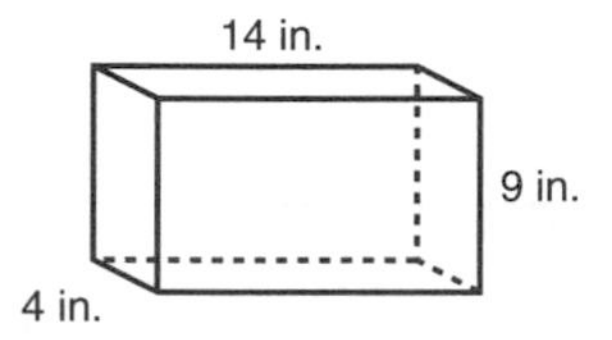

$V =$ __________

5.

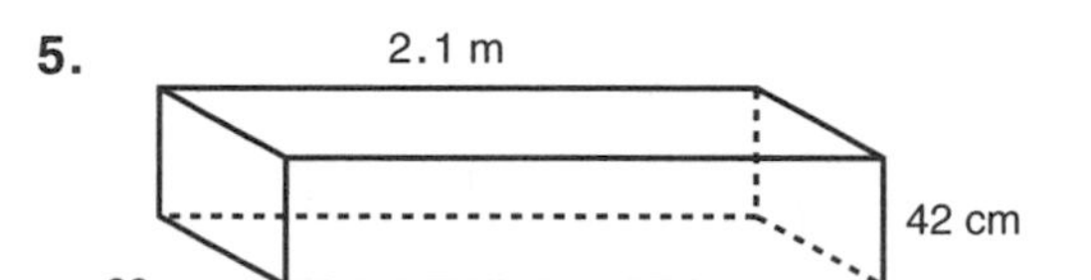

$V =$ __________

Solve.

6. A box is $6\frac{1}{2}$ ft long, $4\frac{1}{4}$ ft wide, and 3 ft high. Find the volume of 7 such boxes. ____________________

7. Find the volume of a rectangular basket 20 cm long, 15.6 cm wide, and 30.4 cm high. ____________________

8. Find, to the nearest tenth, the number of cubic yards of wood there are in a solid mahogany table top $3\frac{1}{2}$ ft long, 2 ft wide, and 3 in. thick. (Hint: 1 cubic yard occupies 27 cubic feet of space.) ____________________

9. What is the volume of a rectangular window box measuring 60 cm long, 24 cm wide, and 20 cm high? ____________________

10. How many cubic yards of air are there in a room 30 yd long, 28 yd wide, and 9 yd high? ____________________

11. At the produce market a vegetable bin 1.6 m long and 90 cm wide is filled to a depth of 76 cm. Find, to the nearest whole number, how many cubic meters of space are used. ____________________

12. A walk-in refrigerator is $6\frac{1}{2}$ ft long, 4 ft wide, and $6\frac{1}{4}$ ft high. About how many cubic yards of space does it occupy? ____________________

Volume: Triangular Prisms*

Name ______________________

Date ______________________

Find the volume of each triangular prism.

$V = Bh; B = \frac{1}{2}bh$

1.

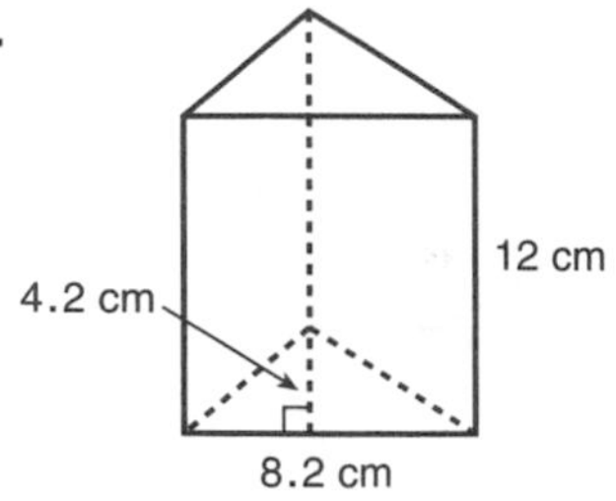

$V =$ ________

2.

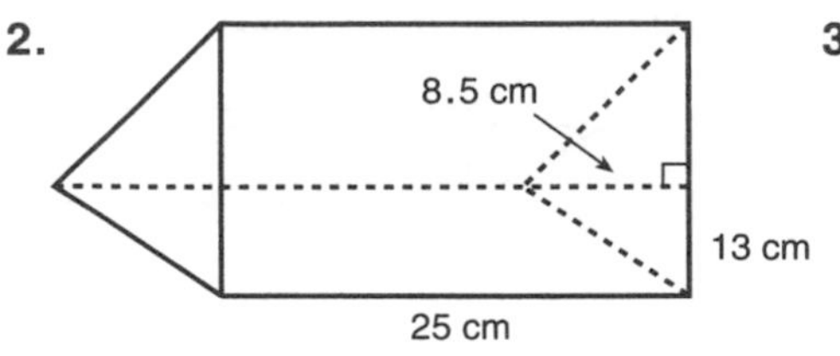

$V =$ ________

3.

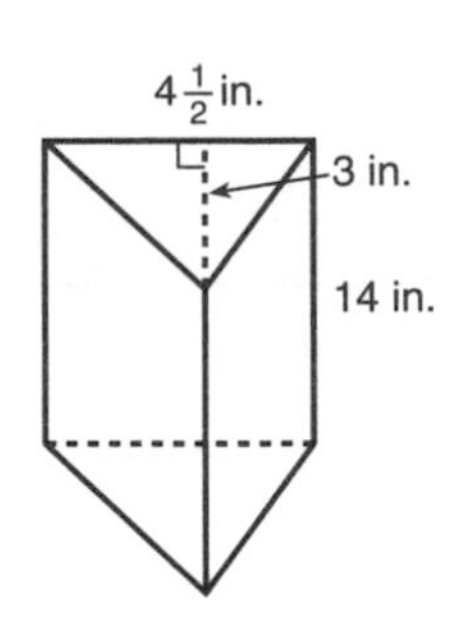

$V =$ ________

4.

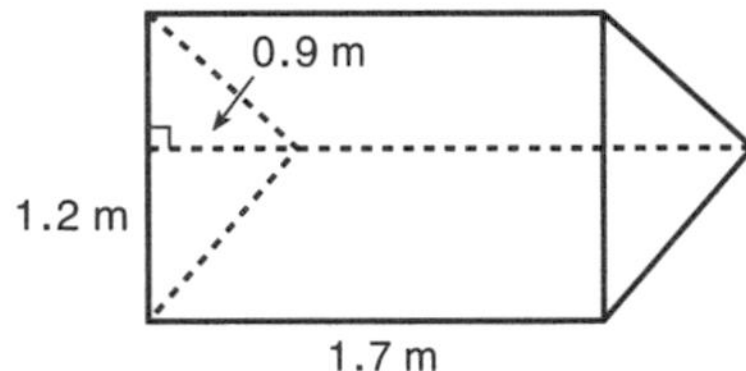

$V =$ ________

5.

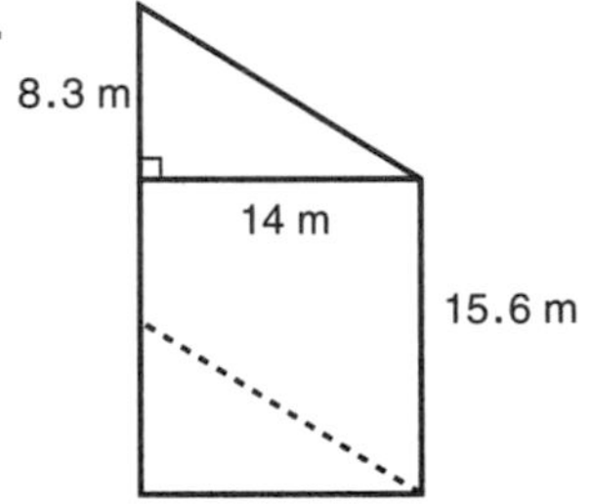

$V =$ ________

6.

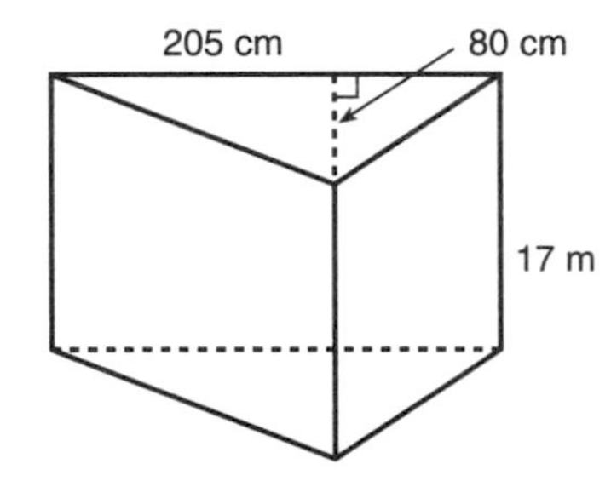

$V =$ ________

Solve.

7. A porch is supported by 6 columns shaped like triangular prisms. Find the amount of concrete used in making these columns if each is 4 yd high, and the area of its base is 15 square yards.

____

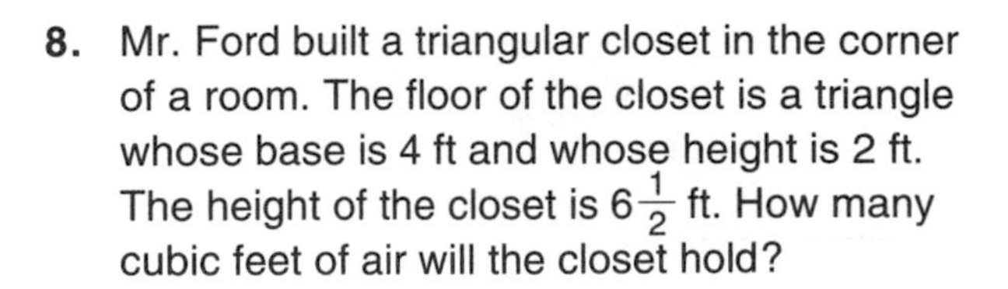

8. Mr. Ford built a triangular closet in the corner of a room. The floor of the closet is a triangle whose base is 4 ft and whose height is 2 ft. The height of the closet is $6\frac{1}{2}$ ft. How many cubic feet of air will the closet hold?

____

9. What is the difference in volume between a rectangular prism whose base measures 3 ft by 4 ft and whose height is 2 ft, and a 3-foot high triangular prism whose base is 6 ft long and 3 ft high?

____

10. Find the volume of the laundry container at the right.

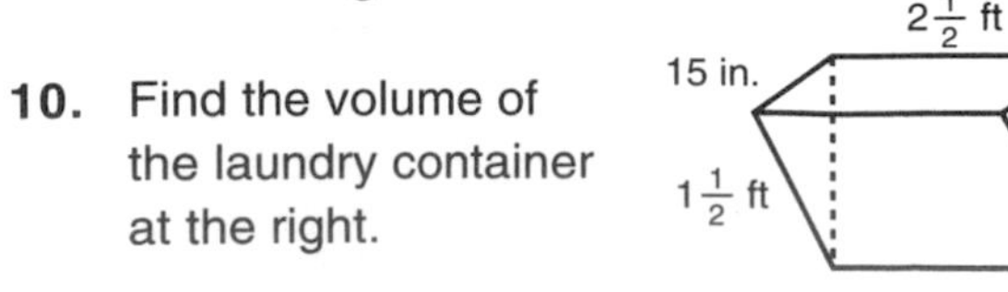

____

*Use with Lesson 13-5, text pages 362–363.

Volume: Pyramids*

Name ______________________

Date ______________________

Find the volume of each pyramid.

$V = \frac{1}{3}Bh$

1.

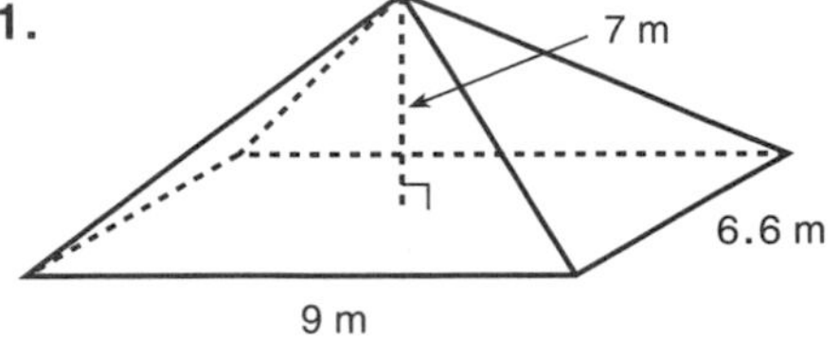

$V =$ __________

2.

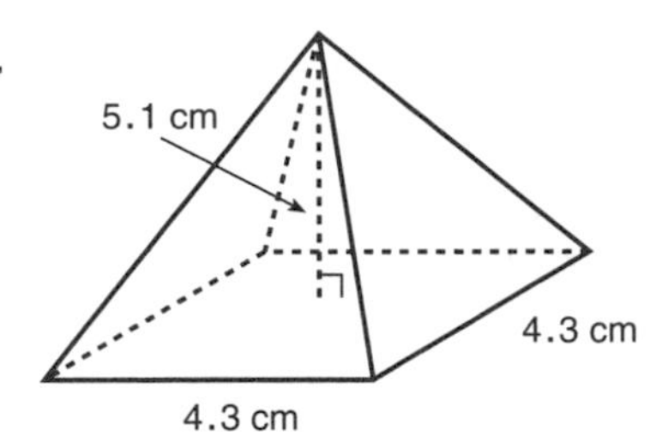

$V =$ __________

3.

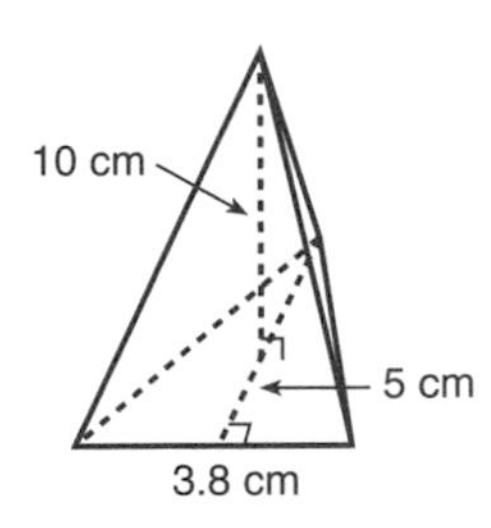

$V =$ __________

4.

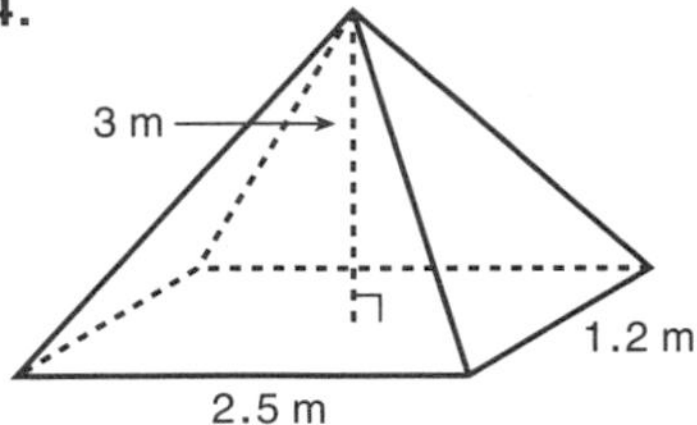

$V =$ __________

5.

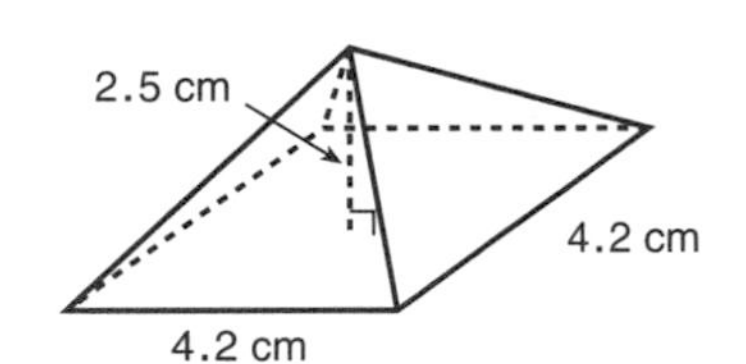

$V =$ __________

6.

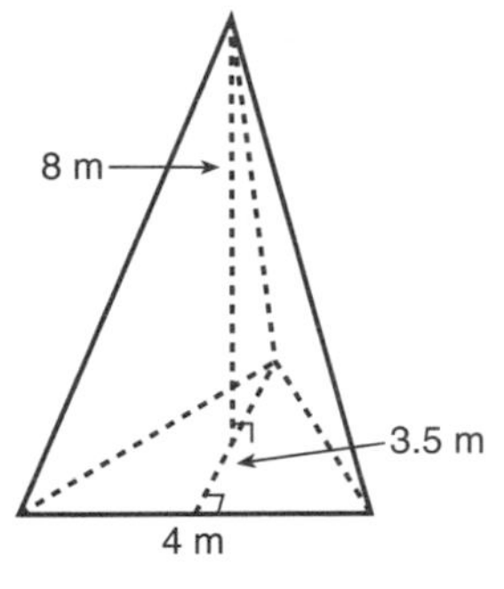

$V =$ __________

Solve.

7. A pyramid has an 8-in. square base. The height of the pyramid is 1 ft. Find the volume. ______________________

8. How many cubic feet of space are occupied by a pyramid-shaped tent with a 9-ft square base and a height of 8 ft? ______________________

9. What is the difference in volume between a square prism with a base measuring 6 cm on an edge and a height of 8.5 cm and a square pyramid with a base measuring 9 cm on an edge and a height of 1 dm? ______________________

10. Find the volume of a square pyramid if the edge of the base measures 10 m and the height is 15 m. ______________________

Volume: Cylinders and Cones*

Name ____________________

Date ____________________

Find the volume. Use 3.14 or $\frac{22}{7}$ for π.

Cylinder: $V = \pi r^2 h$

Cone: $V = \frac{1}{3} \pi r^2 h$

1.

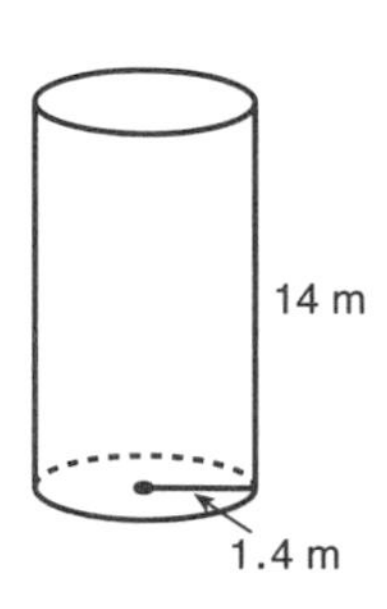

$V \approx$ ________

2.

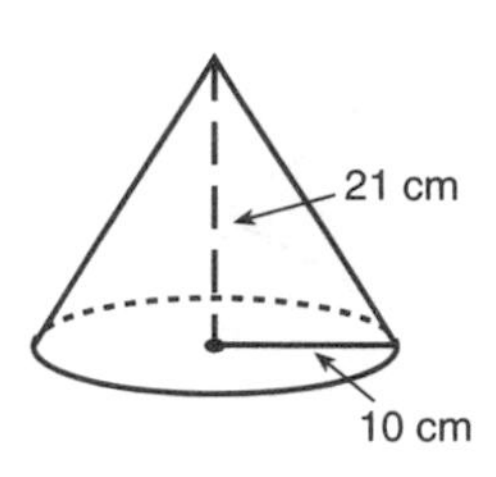

$V \approx$ ________

3.

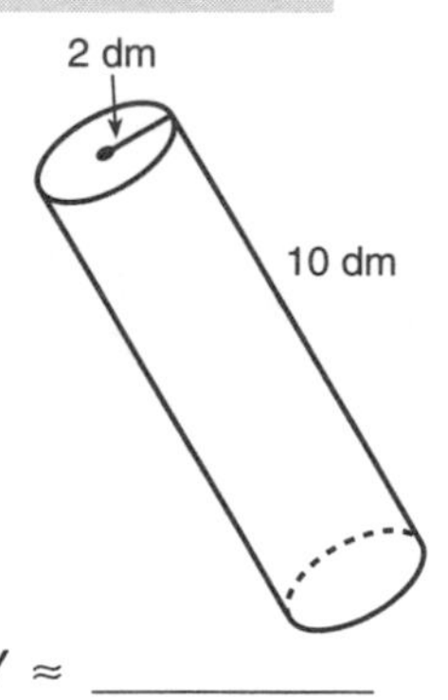

$V \approx$ ________

4.

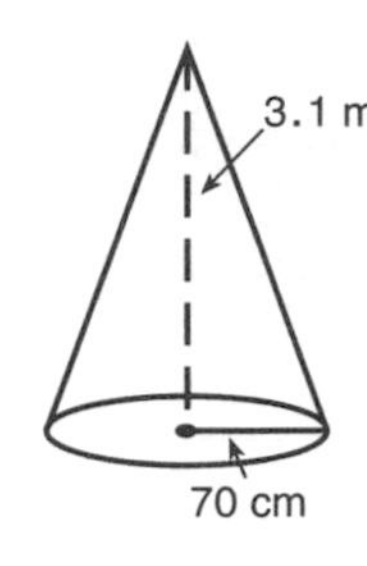

$V \approx$ ________

5.

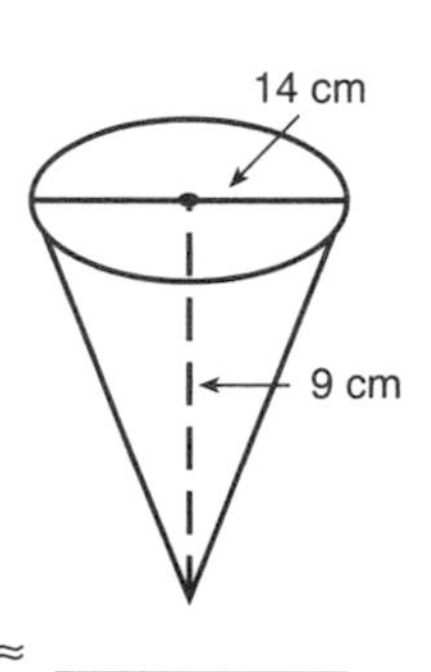

$V \approx$ ________

6.

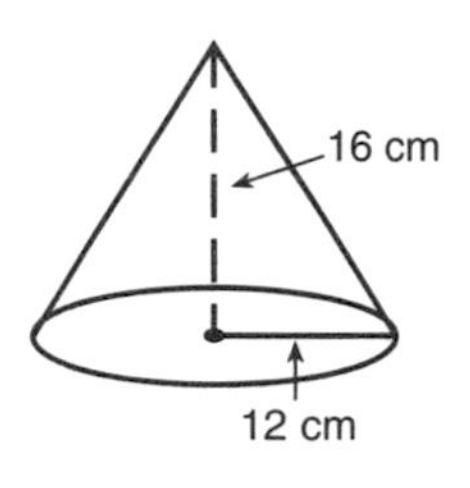

$V \approx$ ________

7.

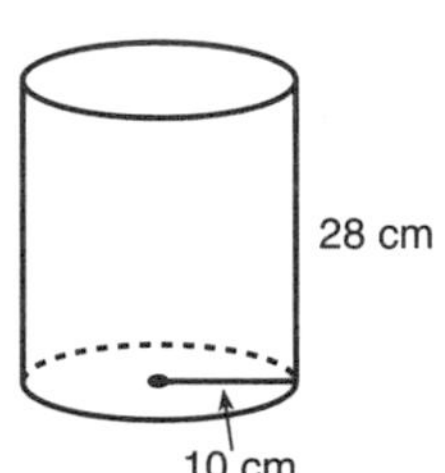

$V \approx$ ________

8.

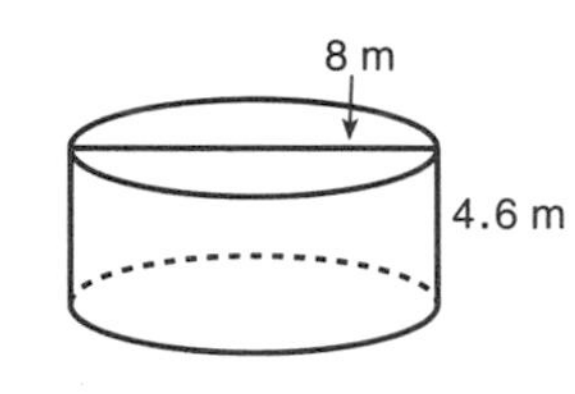

$V \approx$ ________

Solve.

9. What is the volume of a can with a diameter of 6 in. and a height of 10 in.? ____________________

10. How many cubic meters of water will it take to fill a swimming pool that has a radius of 3.5 m and an average depth of 1.5 m? ____________________

11. Estimate the volume of a box of salt that has a radius of 2 in. and a height of $5\frac{1}{2}$ in. ____________________

12. A flower pot has a radius of 8 cm and a depth of 12 cm. How many cubic centimeters of soil will it hold? ____________________

13. Round to the nearest cubic centimeter the total volume of 12 cans of tomatoes. Each can has a base whose diameter is 14 cm and a height of 18 cm. ____________________

14. How much sand will a conical container hold if its base has a diameter of 5 in. and a height of 1 ft 3 in.? ____________________

15. Which holds more, a cylindrical container 6 yd in diameter and 10 yd long or a cubical container 6 yd on an edge? ____________________

*Use with Lesson 13-6, text pages 364–365.

Metric Units: Liquid Volume*

Name ______________________

Date ______________________

Check the best unit of measure for each.

	Item	milliliter	liter	kiloliter
1.	glass of ice tea			
2.	bathtub of water			
3.	water in swimming pool			
4.	tear drop			
5.	oil in tank of a car			
6.	medicine in eye dropper			

Change each to the unit indicated.

7. 8 dm^3 = __________ L

8. 28 m^3 = __________ kL

9. 70 dm^3 = __________ L

10. 25 cm^3 = __________ mL

11. 6 cm^3 = __________ mL

12. 80 m^3 = __________ kL

13. 45 dm^3 = __________ L

14. 18 dm^3 = __________ L

15. 30 cm^3 = __________ mL

16. 3.4 cm^3 = __________ L

17. 18 dm^3 = __________ mL

18. 6280 cm^3 = __________ L

19. 3.4 cm^3 = __________ mL

20. 376 m^3 = __________ L

21. 8756 cm^3 = __________ L

Find the volume.

22.

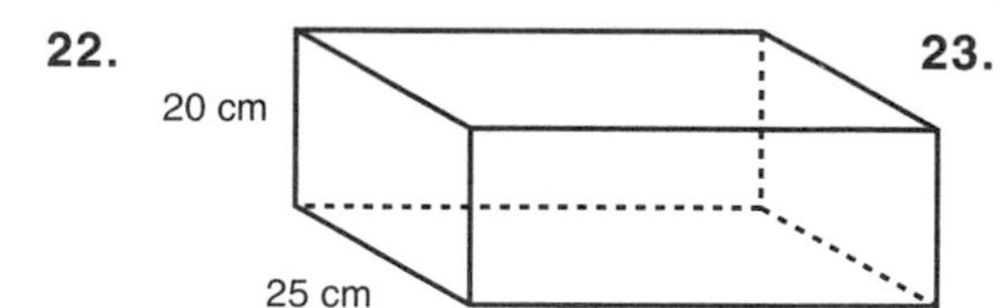

V = __________ mL

23.

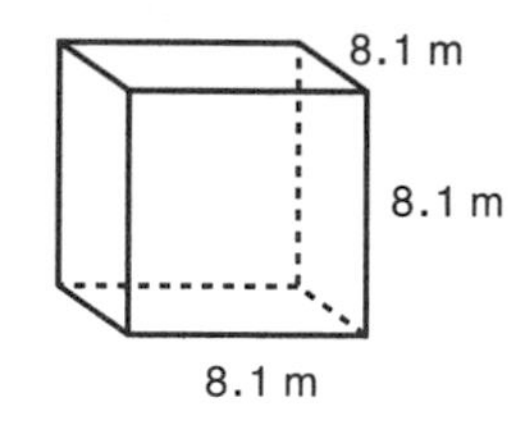

V = __________ kL

24.

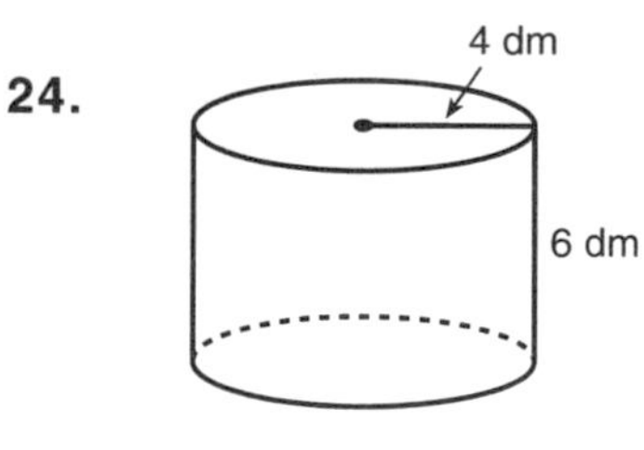

V = __________ L

Solve.

25. How many liters of water will a bucket hold if it is 2.8 dm in diameter and 3 dm in height? (Use $\pi = \frac{22}{7}$.) ______________________

26. How many milliliters of water will fill a cone-shaped paper cup that is 9 cm in diameter and 8 cm high? ______________________

27. A water tank is 8 m in diameter and 15 m deep. How many kiloliters will it hold when it is three fourths filled? ______________________

*Use with Lesson 13-7, text pages 366–367.

Metric Units: Mass*

Name ____________________

Date ____________________

Check the best unit of measure for each.

	Item	grams	milligrams	kilograms	metric tons
1.	box of cereal				
2.	load of coal				
3.	your weight				
4.	raisin				
5.	2 cookies				
6.	dime				
7.	candy bar				
8.	basket of apples				

Complete.

9. 6 dm^3 weighs __________ kg

10. 5 cm^3 weighs __________ g

11. 0.006 cm^3 weighs __________ mg

12. 147 dm^3 weighs __________ kg

13. 25 cm^3 weighs __________ g

14. 14 dm^3 weighs __________ kg

15. 135 dm^3 weighs __________ kg

16. 6000 cm^3 weighs __________ kg

17. 2500 kg weighs __________ t

18. 1000 cm^3 weighs __________ g

19. 0.007 cm^3 weighs __________ mg

20. 5000 kg weighs __________ t

Solve.

21. A rectangular tank is 15 cm long, 10 cm wide, and 12 cm deep. How many kg of water will it hold when filled? ____________________

22. What is the weight of a small block of ice 7 cm long, 5.5 cm wide, and 4 cm high? ____________________

*Use with Lesson 13-8, text pages 368–369.

Metric Units: Mass* (cont'd)

Name ____________________

Date ____________________

Complete each matching exercise with the most reasonable measure.

23. _______ mass of a tennis racquet

24. _______ mass of a golf ball

25. _______ mass of a baseball

26. _______ mass of a bowling ball

27. _______ mass of a Ping-Pong ball

a. 145 g

b. 5 kg

c. 40 g

d. 20 g

e. 350 g

Complete.

28. 9 g = __________ mg

29. 5 kg = __________ g

30. 46 cg = __________ kg

31. 6.8 t = __________ kg

32. 0.025 cg = __________ g

33. 490 cg = __________ g

34. 7.63 g = __________ mg

35. 2.9 kg = __________ g

36. 0.48 mg = __________ g

37. 38 g = __________ cg

38. 470 cg = __________ g

39. 890 000 mg = __________ g

40. 11.5 t = __________ kg

41. 11 300 kg = __________ t

Complete the charts.

	Capacity	Cubic Volume	Mass
42.		2 dm^3	2 kg
43.	2 mL	2 cm^3	
44.	5 mL		5 g
45.	1 kL	1 m^3	
46.		10 cm^3	1 dg

	Capacity	Cubic Volume	Mass
47.	10.5 L		
48.			100 g
49.		35 m^3	
50.	8.9 L		
51.		10 dm^3	

*Use with Lesson 13-8, text pages 368–369.

Graphing on the Real Number Line*

Name ______________________

Date ______________________

Find the solution. Graph it on a real number line.

1. $x + 8 = 5$

2. $x - 7 = {}^-6$

3. $x + 6 = 10$

4. $x - 12 = 8$

5. $6x = {}^-18$

6. $5x = 0$

Graph the solution on a real number line.

7. $x + 6 \geq 4$

8. $x - 3 \geq 2$

9. $x + 5 \leq 7$

10. $2x - 6 \leq {}^-2$

11. $4x > 8$

12. ${}^-5x \leq 20$

Write an equation or inequality for each graph.

13. 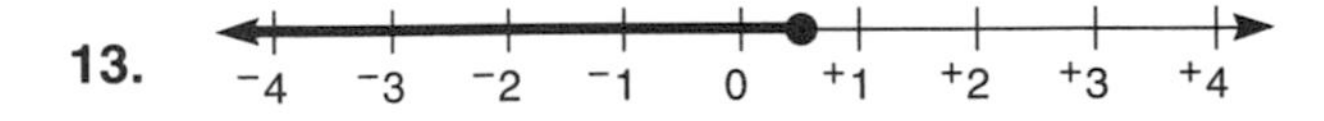

14. 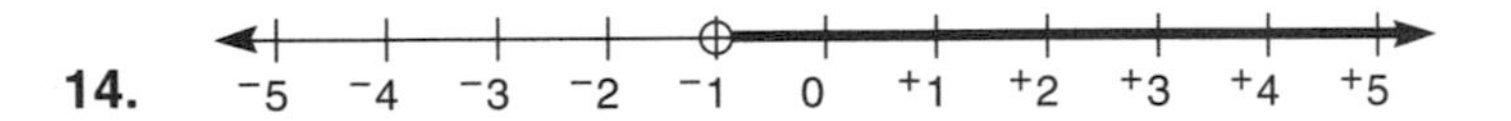

15. 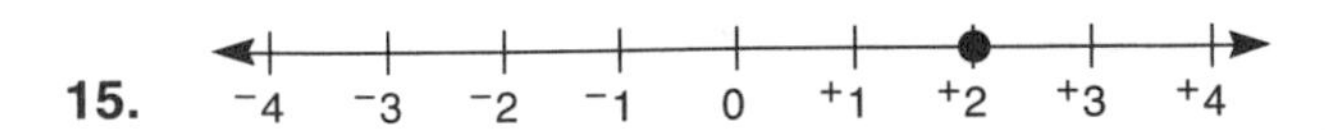

16. 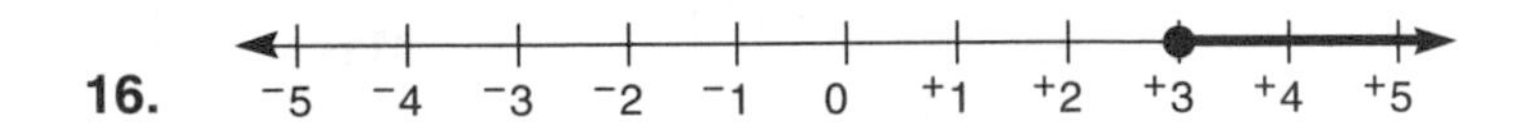

17. 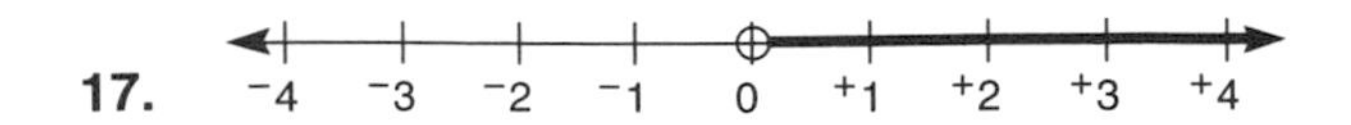

18. ${}^-5 \quad {}^-4 \quad {}^-3 \quad {}^-2 \quad {}^-1 \quad 0 \quad {}^+1 \quad {}^+2 \quad {}^+3 \quad {}^+4 \quad {}^+5$

19. ${}^-4 \quad {}^-3 \quad {}^-2 \quad {}^-1 \quad 0 \quad {}^+1 \quad {}^+2 \quad {}^+3 \quad {}^+4$

20. ${}^-5 \quad {}^-4 \quad {}^-3 \quad {}^-2 \quad {}^-1 \quad 0 \quad {}^+1 \quad {}^+2 \quad {}^+3 \quad {}^+4 \quad {}^+5$

***Use with Lesson 14-1, text pages 380–381.**

Graphing on the Coordinate Plane*
(Four Quadrants)

Name ______________________________

Date ______________________________

Graph the point for each ordered pair.

1. A (2, 6)

2. B ($^{-}4$, 3)

3. C ($^{-}5$, $^{-}2$)

4. D (7, $^{-}5$)

5. E (6, 6)

6. F (0, 0)

7. G ($^{-}4$, $^{-}8$)

8. H ($^{-}8$, 2)

9. I (6, $^{-}1$)

10. J ($^{-}1$, 0)

11. K (0, $^{-}4$)

12. L (2, 0)

13. M (8, $^{-}8$)

14. N (0, 7)

15. O (2, 4)

16. P ($^{-}1$, $^{-}6$)

17. Q (3, $^{-}2$)

18. R ($^{-}5$, 6)

19. S (5, 5)

20. T ($^{-}5$, $^{-}5$)

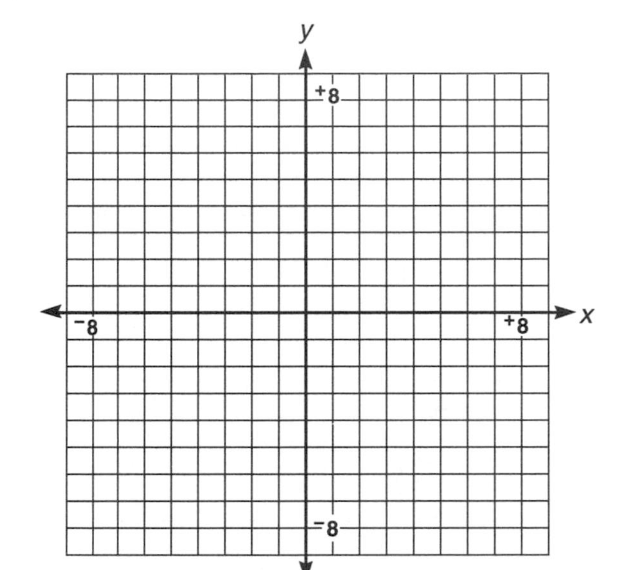

21. Which points are in Quadrant I? ______________________________

22. Which points are in Quadrant II? ______________________________

23. Which points are in Quadrant III? ______________________________

24. Which points are in Quadrant IV? ______________________________

Draw a pair of coordinate axes. Graph these points.

25. A ($^{-}6$, 2)

26. B ($4\frac{1}{2}$, $^{-}2$)

27. C ($^{-}2$, $^{-}5$)

28. D (3, 3)

29. E (1, $^{-}4$)

30. F (0, 3)

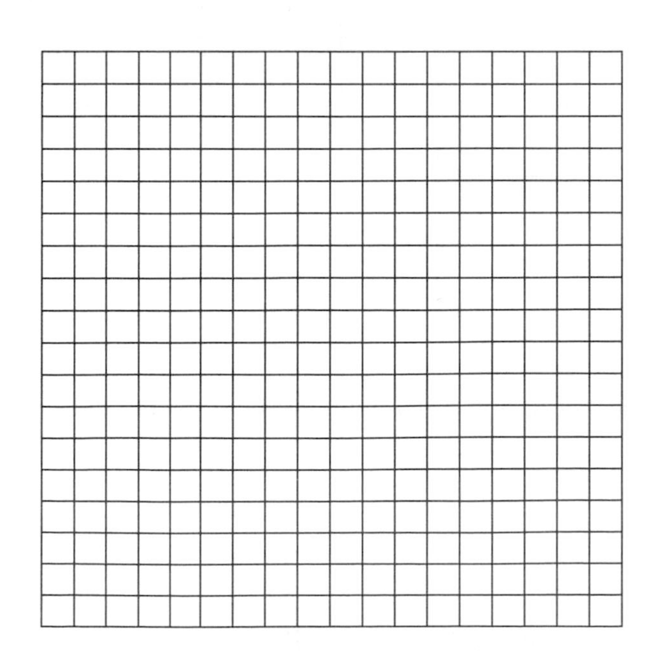

***Use with Lessons 14-2, 14-3, text pages 382–385.**

Equations with Two Variables*

Name ______________________

Date ______________________

Complete each table of values.

1. $y = x + 6$

x	$^-3$	$^-2$	$^-1$	0	1	2	3
y							

2. $y = x - 3$

x	$^-3$	$^-2$	$^-1$	0	1	2	3
y							

3. $y - x = 8$

x	$^-3$	$^-2$	$^-1$	0	1	2	3
y							

4. $y = 2x + 2$

x	$^-3$	$^-2$	$^-1$	0	1	2	3
y							

5. $y = 3x - 1$

x	$^-3$	$^-2$	$^-1$	0	1	2	3
y							

6 $y = 4x - 3$

x	$^-3$	$^-2$	$^-1$	0	1	2	3
y							

Complete each table of values.

	x	$x + 3$	y	ordered pair
7.	$^-2$	$^-2 + 3$		
8.	$^-1$			
9	0			
10.	1			
11.	2			

	x	$3x$	y	ordered pair
12.	$^-2$	$3(^-2)$		
13.	$^-1$			
14	0			
15.	1			
16.	2			

For each equation, find the *y*-value that completes the ordered pair.

17. $y + x = 10$

($^-2$, ____); ($^-1$, ____); (2, ____); (3, ____);

18. $y - 2x = 5$

($^-2$, ____); ($^-1$, ____); (0, ____); (1, ____)

19. $x + y = {}^-6$

($^-1$, ____); (0, ____); (1, ____); (2, ____);

20. $x + 5y = 8$

($^-7$, ____); ($^-2$, ____); (3, ____); (8, ____)

*Use with Lesson 14-4, text pages 386–387.

Graphing Equations*

Name ______________________

Date ______________________

Complete each table and graph the equation on the given coordinate grid.

1. $y = x + 3$

x	$^-2$	$^-1$	0	1	2
y					

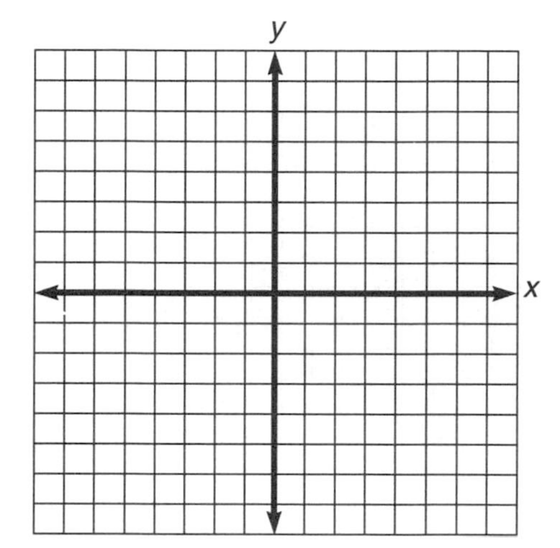

2. $y = 2x - 1$

x	$^-2$	$^-1$	0	1	2
y					

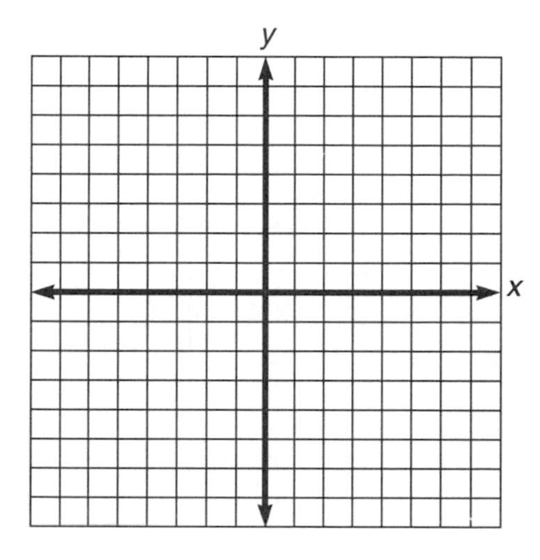

3. $y = 3 - 2x$

x	$^-2$	$^-1$	0	1	2
y					

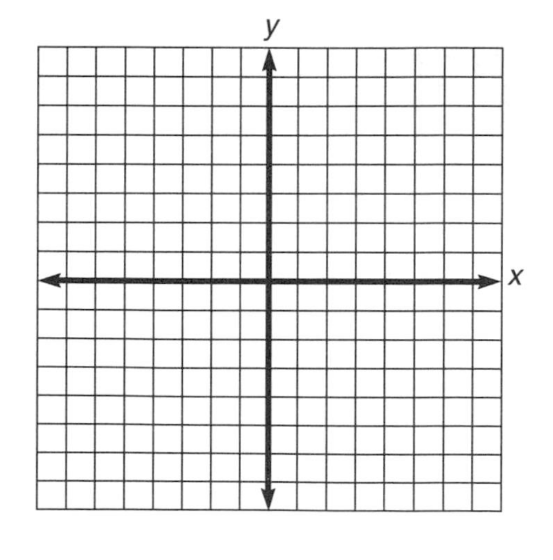

4. $y = \frac{x}{2} + 5$

x	$^-4$	$^-2$	0	2	4
y					

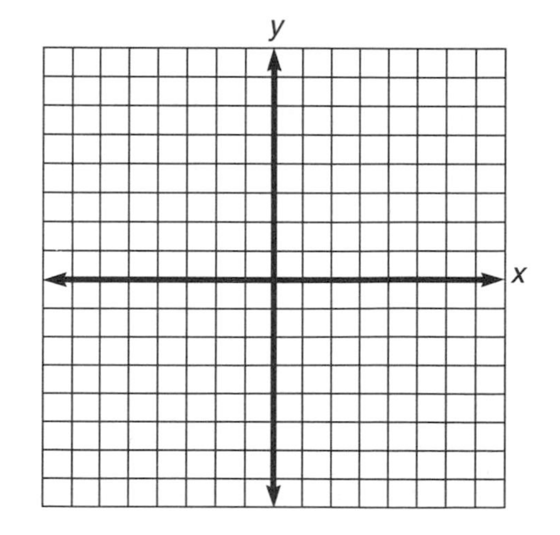

Draw a pair of coordinate axes and graph each equation using 5 points.

5. $y = x - 4$

6. $y = 3 - 2x$

7. $y = \frac{1}{3}x + 2$

7. $y = 7 - \frac{1}{4}x$

*Use with Lesson 14-5, text pages 388–389.

Solving Systems of Equations*

Name ______________________

Date ______________________

Complete each function table. Graph each pair of equations on the same coordinate grid. What is their common solution?

1. $y = x + 2$; $y = 3x$

x	y
−3	
−2	
−1	
0	
1	
2	
3	

x	y
−3	
−2	
−1	
0	
1	
2	
3	

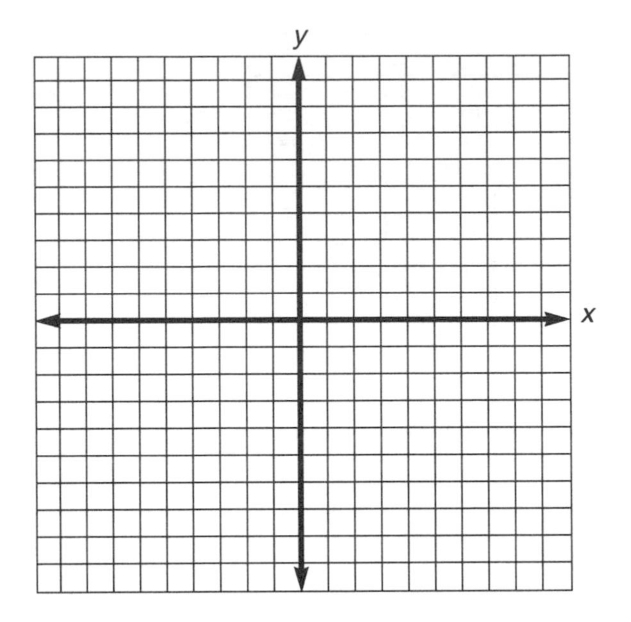

Common Solution: ______________________

2. $y = 3x + 1$; $y = 2x - 1$

x	y
−3	
−1	
$-\frac{1}{3}$	
0	
1	
2	

x	y
−4	
−2	
$-\frac{1}{2}$	
0	
1	
2	

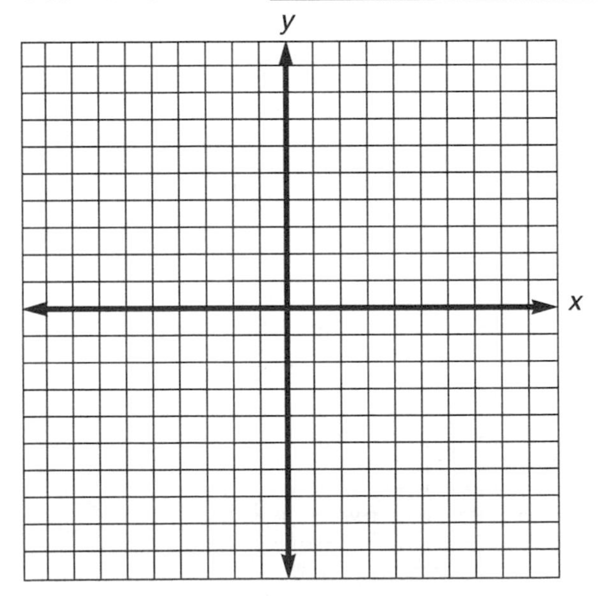

Common Solution: ______________________

Draw a pair of coordinate axes and graph each pair of equations. Where do they intersect? (Use separate paper.)

3. $y = 2 - x$
$y = x + 8$

4. $x + y = 0$
$x - y = 2$

5. $y = \frac{x}{2}$
$y = x + 1$

6. $3y + x = 1$
$x + y = 3$

*Use with Lesson 14-6, text pages 390–391.

Simultaneous Equations*

Name ____________________

Date ____________________

$y = 2x + 2$ and $y - x = 3$ (Remember: $y - x = 3$ is the same as $y = x + 3$)

$x + 3 = 2x + 2$

$\not{x} + 3 - \not{x} - 2 = 2x + \not{2} - x - \not{2}$

$3 - 2 = 2x - x$

$1 = x$

$y - 1 = 3$

$y = 4$

Solution: (1, 4)

Solve the equations by substitution and find the common solution.

1. $y = 6x$
   $y = x - 10$

2. $y = {}^{-}7x$
   $y = x + 8$

3. $y = \frac{x}{4}$
   $y = x + 3$

4. $y = \frac{-x}{5}$
   $y = x - 6$

5. $y = x + 5$
   $y = 4x - 1$

6. $y = {}^{-}x + 4$
   $y = 2x - 5$

7. $y = {}^{-}5x + 2$
   $y = 3x - 14$

8. $x + y = 5$
   $2x - 3y = 5$

9. $x + y = 0$
   $5x + y = 4$

Solve the equations by addition or subtraction. Then find the common solution.

10. $x + y = 12$
    $\underline{x - y = 8}$

11. $3x + y = 7$
    $\underline{{}^{-}3x + y = 13}$

12. $4x - y = 3$
    $\underline{2x - y = 7}$

13. $2x + 3y = 2$
    $\underline{2x - y = {}^{-}6}$

14. $x - y = 6$
    $\underline{2x - y = 4}$

15. $6x - 2y = 36$
    $\underline{5x - 2y = 25}$

*Use with Lesson 14-7, text pages 392–393.

Graphing Inequalities*

Name ______________________

Date ______________________

Write y in terms of x. Describe each boundary line.

1. $3x - y < 5$ ______________
2. $x - y < 4$ ______________
3. $2x + y < 3$ ______________
4. $6x - 4y > 8$ ______________
5. $4x - 6y \geq 2$ ______________
6. $4x + 4y \leq 12$ ______________
7. $x + 2y \geq 6$ ______________
8. $5x - y > {}^{-}3$ ______________

Choose test points to solve each graph. Shade the correct half-plane.

9. $x + y \geq 2$

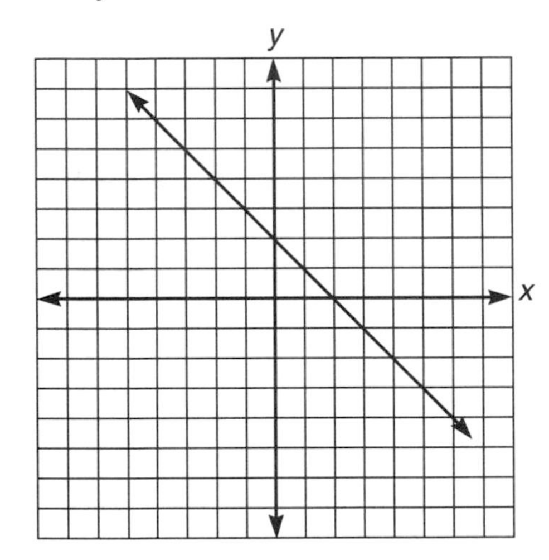

10. $y < \frac{1}{2}x + 3$

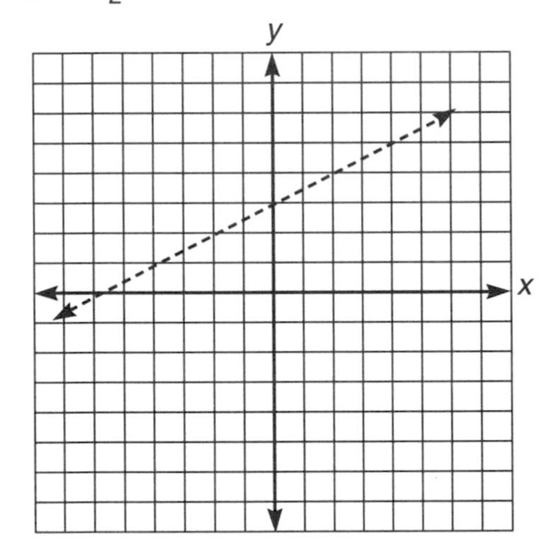

Describe each boundary line. Graph the solution set.

11. $3x + y \geq 4$

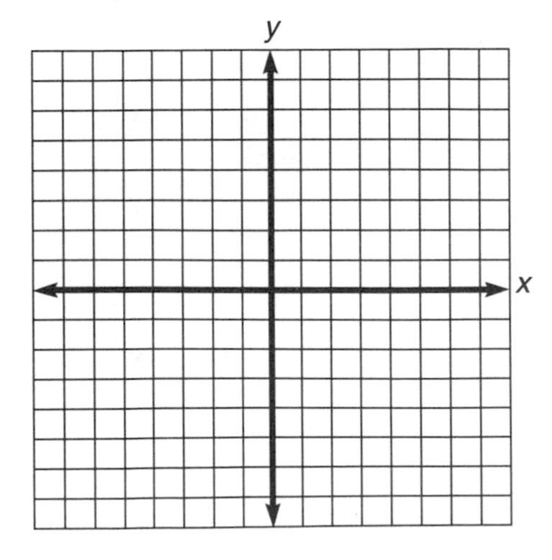

12. $2x - 5y > 5$

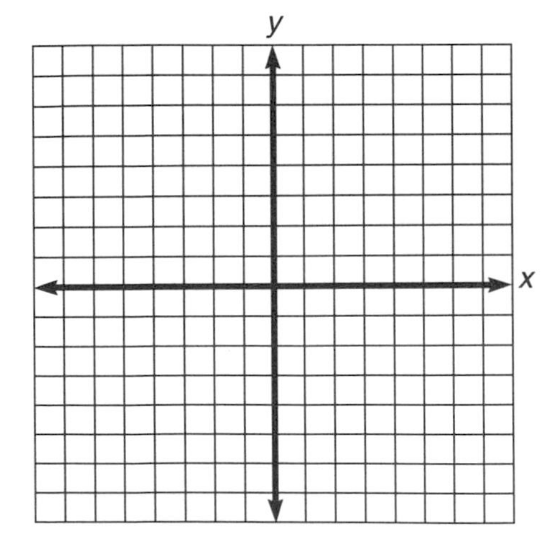

*Use with Lesson 14-8, text pages 394–395.

Graphing Systems of Inequalities*

Name ______________________

Date ______________________

Which point is part of the solution set of the given system of inequalities? Circle the correct answer.

1. $x + y \leq 4$

$x + y \geq {}^{-}2$

a. (0, 0) **b.** (3, 3) **c.** (⁻2, ⁻2) **d.** (⁻2, ⁻3)

2. $y - 2x \geq 0$

$y - x > 2$

a. (0, 0) **b.** (1, 1) **c.** (⁻4, ⁻1) **d.** (4, ⁻1)

3. $x - y \leq 1$

$2x - y \geq 2$

a. (0, 0) **b.** (2, 2) **c.** (⁻3, ⁻2) **d.** (⁻2, 3)

Graph the solution set of each system of inequalities. Use the coordinate grid.

4. $x < 0$
$x + 2y < 2$

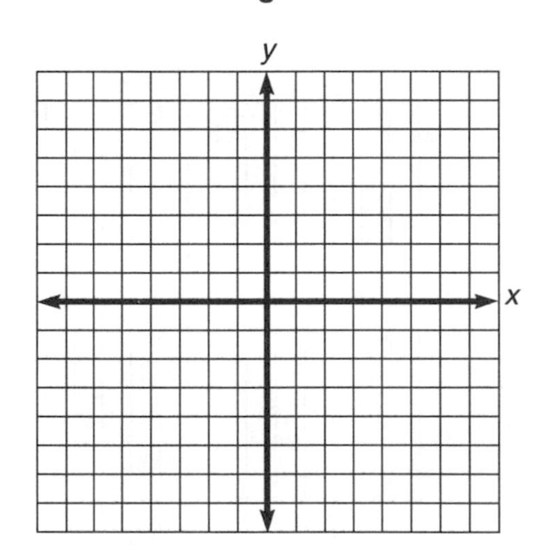

5. $y > {}^{-}2x$
$y - 2x > 3$

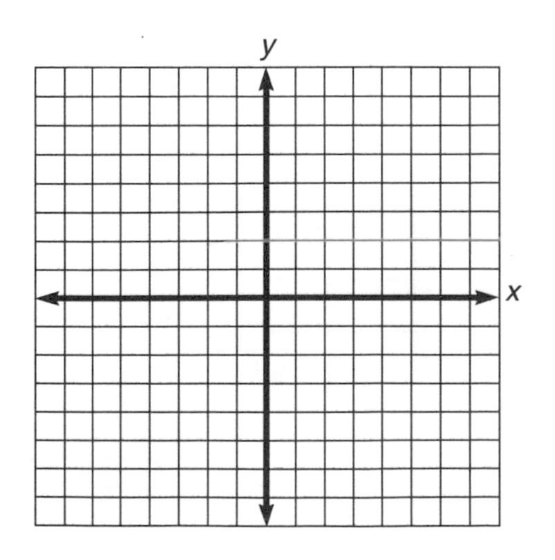

*Use with Lesson 14-9, text pages 396–397.

Graphing a Curve*

Name ______________________

Date ______________________

Complete each function table and graph the ordered pairs. Connect the points with a smooth curve.

1. $y = x^2 + 1$

x	y
⁻3	
⁻2	
⁻1	
0	
1	
2	
3	

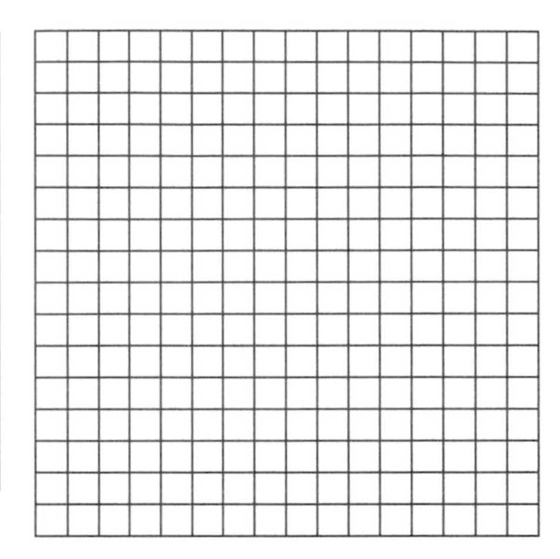

2. $y = x^2 + 2$

x	y
⁻3	
⁻2	
⁻1	
0	
1	
2	
3	

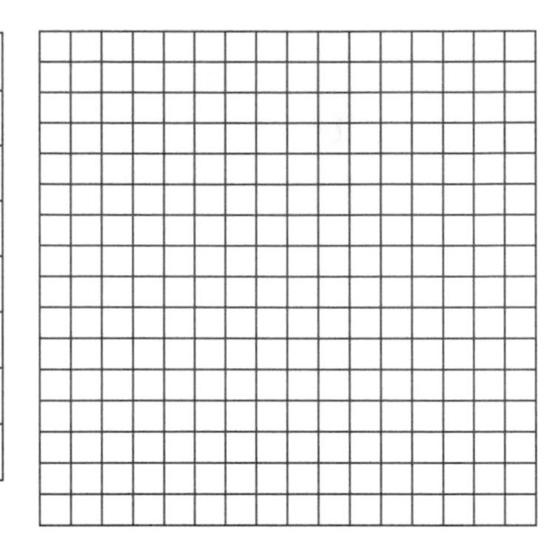

3. $y = x^2 - 2$

x	y
⁻3	
⁻2	
⁻1	
0	
1	
2	
3	

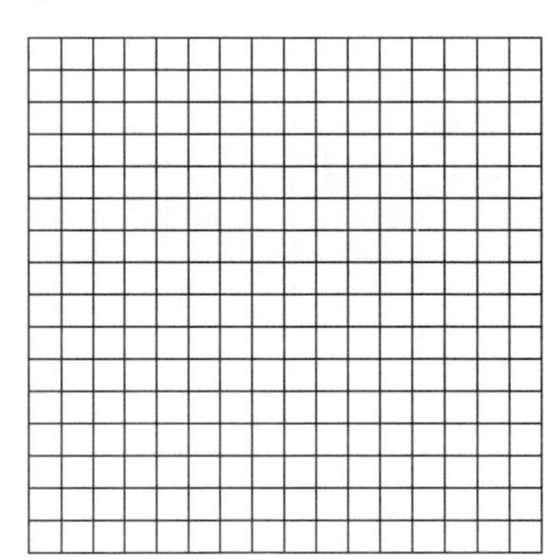

4. $y = 2x^2 - 1$

x	y
⁻3	
⁻2	
⁻1	
0	
1	
2	
3	

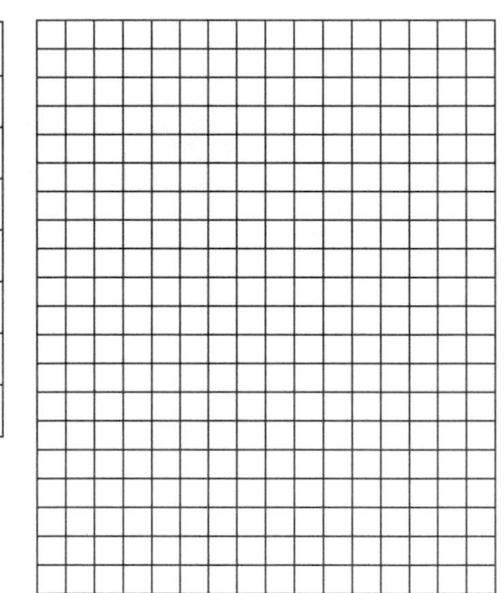

5. $y = {}^{-}(x^2)$

x	y
⁻3	
⁻2	
⁻1	
0	
1	
2	
3	

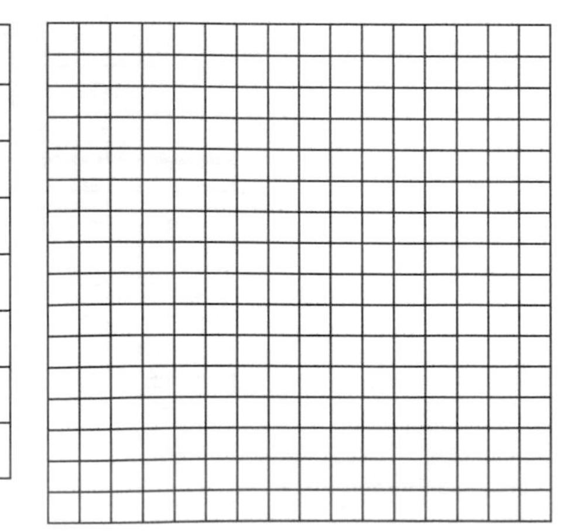

***Use with Lesson 14-10, text pages 398–399.**

Progress Test: Chapters 8–14*

Name ____________________

Date ____________________

Write each answer on the numbered line. Do all computations on a separate sheet of paper.

Choose the correct answer. Write the letter.

Answers

1. A profit of 9.2% on $350 is:
a. $.32 **b.** $3.22 **c.** $32.20 **d.** $322

1. ______ (8–1)

2. A new clock has a list price of $38 and a discount of 15%. How much will it cost?
a. $5.70 **b.** $32.30 **c.** $33.30 **d.** $42.70

2. ______ (8–2)

3. A sofa bed is marked $650 plus tax. If the sales tax is $7\frac{1}{2}$%, what is the total cost?
a. $698.75 **b.** $698.25 **c.** $697.75 **d.** $697.25

3. ______ (8–4)

4. Find the rate of commission on a total sales of $3600 and a commission of $9.
a. 40% **b.** 0.25% **c.** 25% **d.** 2.5%

4. ______ (8–5)

5. The number of degrees in 75% of a circle is:
a. 360° **b.** 270° **c.** 180° **d.** 90°

5. ______ (9–10)

6. The mean for the test scores 76, 89, 83, 90, 92, 80 is:
a. 82 **b.** 84 **c.** 85 **d.** 90

6. ______ (9–11)

7. The probability of getting a number that is a multiple of 6 when tossing a number cube whose faces are marked 3, 6, 9, 12, 15, 18 is:
a. $\frac{1}{2}$ **b.** $\frac{1}{3}$ **c.** $\frac{1}{4}$ **d.** $\frac{1}{6}$

7. ______ (9–1)

8. $\frac{6!}{(6-2)!} = \underline{\ ?\ }$ **a.** 6 **b.** 12 **c.** 24 **d.** 30

8. ______ (9–3)

9. Which of the following is *true*?
a. A line contains only 2 points.
b. Parallel lines intersect.
c. Ray ST is the same as ray TS.
d. ∠PRT can also be called ∠TRP.

9. ______ (10–1)

10. Which of the following is *false*?
a. Two right angles are supplementary.
b. Two acute angles may be complementary.
c. Angles that have a common vertex are vertical.
d. Adjacent angles have a common side and common vertex.

10. ______ (10–4)

11. What is the length of a rectangle whose perimeter is 64 cm if the length is 3 times the width?
a. 24 cm **b.** 18 cm **c.** 16 cm **d.** 8 cm

11. ______ (11–1)

12. Find the side of a square whose area is the same as that of a rectangle 7 inches wide and 28 inches long.
a. 7 in. **b.** 14 in. **c.** 16 in. **d.** 18 in.

12. ______ (11–5)

13. The radius of a circle whose area is equal to its circumference is:
a. 1 **b.** 6 **c.** 2 **d.** 4

13. ______ (11–5)

***Next to each item is given the lesson number in the text where the item was taught.**

Progress Test: Chapters 8–14*

Name ____________________

Date ____________________

14. The area of a trapezoid whose altitude is 5 ft and bases are 8 ft and 6 ft is:
a. 70 sq ft **b.** 50 sq ft **c.** 45 sq ft **d.** 35 sq ft
14. __________ (11–4)

15. $0.\overline{7} = \underline{\ ?\ }$ **a.** $\frac{7}{11}$ **b.** $\frac{7}{8}$ **c.** $\frac{7}{9}$ **d.** $\frac{7}{10}$
15. __________ (12–1)

16. $\sqrt{200}$ would be between which two whole numbers?
a. 14 and 15 **b.** 13 and 14 **c.** 12 and 13 **d.** 10 and 11
16. __________ (12–2)

17. The square of a number is 0.64. Find the number.
a. 0.08 **b.** 0.8 **c.** 8 **d.** 16
17. __________ (12–3)

18. The surface area of a cube with an edge of 8 cm is:
a. 256 cm^2 **b.** 384 cm^2 **c.** 192 cm^2 **d.** 64 cm^2
18. __________ (13–1)

19. The volume of a cube with an edge of 6 in. is:
a. 196 $in.^3$ **b.** 392 $in.^3$ **c.** 108 $in.^3$ **d.** 216 $in.^3$
19. __________ (13–4)

20. The volume of a cylinder that has an altitude of 5 ft and the radius of its base equal to 7 ft is approximately:
a. 850 ft^3 **b.** 840 ft^3 **c.** 770 ft^3 **d.** 720 ft^3
20. __________ (13–6)

21. Which point satisfies the equation $y = x - 2$?
a. (2, 1) **b.** (2, $^{-}4$) **c.** (3, $^{-}5$) **d.** (4, 2)
21. __________ (14–4)

22. Which point lies in Quadrant II?
a. ($^{-}3$, 0) **b.** (6, $^{-}2$) **c.** ($^{-}6$, 2) **d.** (3, 0)
22. __________ (14–3)

23. The given graph

is the solution of which inquality?
a. $x < {}^{-}2$ **b.** $x \geq {}^{-}2$ **c.** $x > {}^{-}2$ **d.** $x \leq {}^{-}2$
23. __________ (14–1)

24. Write y in terms of x for the equation $4x + 3y = 7$.
a. $y = {}^{-}\frac{4}{3}x + \frac{7}{3}$ **b.** $y = {}^{-}\frac{4}{3}x + 7$
c. $y = {}^{-}4x + \frac{7}{3}$ **d.** $y = {}^{-}4x + 7$
24. __________ (14–7)

Solve.

25. Find the interest on \$750 for 5 years at $5\frac{1}{2}$%.
25. __________ (8–7)

26. By selling a painting valued at \$150 for \$195, an art museum realized what rate of gain?
26. __________ (8–6)

27. One of the angles of a right triangle measures 53°. What are the measures of the other two angles?
27. __________ (10–5)

28. Find the height of a triangle if its area is 68.8 cm^2 and its base is 16 cm.
28. __________ (11–3)

29. What is the surface area of 6 spherical candles each having a diameter of 7 inches?
29. __________ (13–3)

30. The sum of two numbers is 66. The larger number is 4 more than the smaller number. Find the numbers.
30. __________ (14–13)

*Next to each item is given the lesson number in the text where the item was taught.

Classifying and Evaluating Polynomials*

Name ______________________

Date ______________________

Classify each polynomial as a *monomial (M), binomial (B)* or *trinomial (T)*. Give its degree.

1. $-3x^3y + 5x$ ________
2. $5a + 1$ ________
3. $9x^2y + 4x^3 + 3xy^2$ ________
4. 21 ________
5. $-7m^3n^2$ ________
6. $2b^2 - 3b + 1$ ________
7. -10 ________
8. $\frac{1}{2}a^2b + \frac{3}{4}ab^2$ ________
9. $-7x + 9xy + 10xyz$ ________
10. $\sqrt{2}x^4y^2 - \sqrt{3}x^3y^3 + \sqrt{5}x^5y^2$ ________
11. $1.32mn^2 - 0.35m^3 + 2.43m^4n^2$ ________

Evaluate each expression for $x = -3$ and $y = 4$.

12. $-x^3$ ________
13. $(-x)^3$ ________
14. $x^2 + 6xy - 7y^2$ ________

Evaluate each expression for $a = \frac{1}{3}$, $b = \frac{2}{9}$, and $c = \frac{-1}{6}$.

15. abc ________
16. $a^2c - 6b$ ________
17. $a^3 + bc$ ________

For the following geometric figures find the measures represented by the algebraic expressions, given $a = 3$ and $b = 7$. Use these results to find the area of each figure.

18. a square: length of a side $= a + b$. ________
19. a rectangle: length $= 3a + 2b$; width $= 5a - b$. ________
20. a triangle: base $= 5a + 2$; height $= 3b - a$. ________
21. a trapezoid: height $= \dfrac{3a^2 - 7}{5}$;
    bases $= b + 2$; $2a - 1$. ________
22. a circle: radius $= \frac{1}{3}a + 7$. ________

*Use with Lesson 15-1, text pages 416–417.

Classifying and Evaluating Polynomials* (con't)

Name ______________________

Date ______________________

Find the volume of each geometric figure, given that the measures of its parts are the algebraic expressions and that $a = 2$, $b = 5$.

23. cube: length of an edge $= 2(5a - b)$ ______________________

24. cube: length of an edge $= 3a^2 - b$ ______________________

25. rectangular prism: length $= a + b$,
width $= b - 2$, height $= 2a$ ______________________

26. rectangular pyramid: length of base $= \frac{1}{2}(7a + 2b)$;
width of base $= \frac{1}{3}(4b - a)$
height $= \frac{1}{5}(10a + b)$ ______________________

27. cylinder: radius of base $= 3b - 4a$; height $= 6a + b$ ______________________

Choose the values of *m* and *n* that will make each equation true.

28. $m^2 + n^2 = 5$

a.	$m = 4$ $n = {}^-1$	**b.**	$m = {}^-2$ $n = 1$	**c.**	$m = 3$ $n = 2$	**d.**	$m = {}^-3$ $n = {}^-2$

____

29. $mn + m^2n^2 = 6$

a.	$m = 0$ $n = {}^-6$	**b.**	$m = 6$ $n = 1$	**c.**	$m = 1$ $n = {}^-3$	**d.**	$m = {}^-1$ $n = {}^-3$

____

30. $m + 3n + 2 = 0$

a.	$m = 1$ $n = {}^-1$	**b.**	$m = 0$ $n = 0$	**c.**	$m = {}^-1$ $n = 1$	**d.**	$m = 2$ $n = {}^-2$

____

31. $m^2 - 6mn + 9 = -23$

a.	$m = 3$ $n = 2$	**b.**	$m = {}^-3$ $n = {}^-2$	**c.**	$m = {}^-2$ $n = 3$	**d.**	$m = 2$ $n = 3$

____

*Use with Lesson 15-1, text pages 416–417.

Addition and Subtraction of Polynomials*

Name ______________________

Date ______________________

$$9a - 3b + 5a = \underline{?}$$

$$\begin{array}{r} 9a - 3b \\ +5a \\ \hline 14a - 3b \end{array}$$

$$(4x^2 + 3xy - 9y^2) - (5xy - 4y^2) = \underline{?}$$

$$\begin{array}{r} 4x^2 + 3xy - 9y^2 \\ \overset{-}{\cancel{+}}\ 5xy\ \overset{+}{\cancel{-}}\ 4y^2 \\ \hline 4x^2 - 2xy - 5y^2 \end{array}$$

Add or subtract.

1. $16x^2y^3 + 19x^2y^3$ ______________________

2. $(16ab + c) + (12ab - 4c)$ ______________________

3. $(17c - 4d) - (11c + 4d)$ ______________________

4. $(x^2 - 23x + 14) + (6x^2 - 8)$ ______________________

5. $(\frac{1}{4}pq - \frac{2}{3}cd) + (\frac{3}{8}pq - \frac{4}{9}cd)$ ______________________

6. $(^-\frac{5}{6}rs + \frac{7}{11}m^2n^2) - (^-\frac{2}{3}rs + \frac{3}{22}m^2n^2)$ ______________________

7. $(2b^2 - 1.3n^2) - (3.5n^2 + b^2)$ ______________________

8. $(3rs + 1.5s^2) - (2.3s - 5rs)$ ______________________

Add.

9. $\begin{array}{r} -23a^2 + 17a + 6 \\ 18a^2 + 11a - 7 \\ \hline \end{array}$

10. $\begin{array}{rr} 8xy & + 1 \\ -xy - 10y^2 & \\ \hline \end{array}$

11. $\begin{array}{rr} 47m^2 - 36mn & + 11n^2 \\ 16m^2 & - 9n^2 \\ \hline \end{array}$

12. $\begin{array}{r} 11.3x^2 - 13.9x + 15.8 \\ -9.6x^2 + 12.7x - 6.5 \\ \hline \end{array}$

13. $\begin{array}{r} 10d^2 - 3d \\ 7d^2 + 3d - 9 \\ \hline \end{array}$

14. $\begin{array}{r} 57.6c^2d^2 + 19.7cd - 64.5 \\ 16.1cd - 11.3 \\ \hline \end{array}$

Subtract.

15. $\begin{array}{r} 98a^2 - 121b^2 \\ 24a^2 - 35b^2 \\ \hline \end{array}$

16. $\begin{array}{r} 11c^3 - 13 \\ 7c^3 - 15 \\ \hline \end{array}$

17. $\begin{array}{l} 125x^2y + 57xy^2 - 13xy \\ \quad 90x^2y - 13xy^2 \\ \hline \end{array}$

18. $\begin{array}{r} 16r - 12p \\ 27p + 19 \\ \hline \end{array}$

19. $\begin{array}{l} -9xy \\ -9xy - y^2 \\ \hline \end{array}$

20. $\begin{array}{r} 96.7m^2n^2 + 13.7mn - 105.1n^2 \\ 14.3m^2n^2 - 11.3mn - 13.9n^2 \\ \hline \end{array}$

*Use with Lesson 15-2, text pages 418–419.

Addition and Subtraction of Polynomials* (con't)

Name ____________________

Date ____________________

Write an algebraic expression to represent the perimeter of each geometric figure.

21. square: length of a side $= 5x - 2$ ____________________

22. square: length of a side $= x^2 - 2x + 3$ ____________________

23. rectangle: length $= 9x + 1$; width $= 3x - 2$ ____________________

24. rectangle: length $= 3x^2 - 4x + 1$; width $= 8x + 3$ ____________________

Arrange in ascending powers of *a*.

25. $a + 10 - 3a^2$

____

26. $5a^3 + a - 6a^2 + 11$

____

27. $4ab + 10 - 7a^2b^2 + 9a^4b^3$

____

Arrange in descending powers of *x*.

28. $10x^5 + 9x^3 + 3 - x^4 + x^2$

____

29. $x^3y^2 - 15x^5 + 3 - x^4y + x^2y$

____

30. $-x^4y + 10y + x^8 + 4x^3y^2 - x^6y^5$

____

Solve. Show your work.

31. The sum of two polynomials is $8a^2 - 7$. If one of the polynomials is $3a^2 - 2b - 1$, what is the other polynomial? ____________________

32. From the sum of $3a^2 - 3b$ and $7 - 11b$ subtract $3a^2 - 17$. ____________________

33. The lengths of the sides of a triangle are represented by $x^2 + 8x - 17$, $3x + 11$, and $3x^2 - 25$. Find the perimeter. ____________________

34. The length of each side of a square is $\frac{2a}{3} + 1$. Find the perimeter of the square. ____________________

*Use with Lesson 15-2, text pages 418–419.

Exponents*

Name ____________________

Date ____________________

Simplify using the law(s) of exponents.

1. $a^4 \cdot a^7$ ________

2. $y^3 \cdot y^{10}$ ________

3. $n^2 \cdot n^4 \cdot n$ ________

4. $b^4 \cdot b^2 \cdot b^3$ ________

5. $\frac{x^{12}}{x^7}$ ________

6. $\frac{y^{17}}{y^{25}}$ ________

7. $\frac{m^{14}}{m^{14}}$ ________

8. $\frac{c^{11}}{c^{12}}$ ________

9. $(m^3)^5$ ________

10. $(n^7)^4$ ________

11. $(d^6)^6$ ________

12. $(e^8)^9$ ________

13. $(xy)^7$ ________

14. $(2abc)^3$ ________

15. $(3x^2y^3)^4$ ________

16. $(5mn^5)^2$ ________

17. $(10a^3b^2)^3 \cdot (ab^3)^2$ ________

18. $(-xyz^3)^2 \cdot (-xy)^6$ ________

19. $(a^4b^5c^6)^2$ ________

20. $(m^3n^5y^2)^4$ ________

21. $\frac{(x^4y^2z^3)^4}{(xyz)^2}$ ________

22. $\frac{(x^2y^7z^3)^5}{x^{10}y^7}$ ________

23. $(2x^4yz)^5$ ________

24. $(-2a^5y^4z^8)^3$ ________

Find the value of each of the following:

25. $\frac{3^6 \cdot 3^3}{3^5}$ ________

26. $\frac{8^4 \cdot 8^9}{8^3 \cdot 8^8}$ ________

27. $\frac{4^3 \cdot 4^0}{4^2 \cdot 4^3}$ ________

28. $\frac{9^{12} \cdot 9^3}{(9^3)^5}$ ________

29. $\frac{7^8}{7^5 \cdot 7}$ ________

30. $\frac{10^6}{(10^3)^3}$ ________

*Use with Lesson 15-3, text page 420.

Multiplication of Monomials*

Name ______________________

Date ______________________

$(-3b^2)(7b^5) = \underline{\ ?\ }$
$(-3)(7)(b^2)(b^5) =$
$-21b^{2+5} = -21b^7$

$2y(3y^2 + 9) = \underline{\ ?\ }$
$(2y \cdot 3y^2) + (2y \cdot 9) = 6y^{1+2} + 18y$
$= 6y^3 + 18y$

Multiply.

1. $15x^4 \cdot 4x^7$ ______
2. $-13y^5 \cdot 9y^{10}$ ______
3. $(103a^2)(-2a^4)$ ______
4. $(-25b^6)(-7b^3)$ ______
5. $(16x^2y^3)(4x^7y^9)$ ______
6. $(-17m^4b^7c)(8m^3c^5)$ ______
7. $5c^4 \cdot -8c^5 \cdot 2c^3$ ______
8. $-11a^5 \cdot -2a^6 \cdot 4a^3$ ______
9. $\left(\frac{2}{3}a^5b^4c^2\right)^3$ ______
10. $\left(-\frac{3}{5}e^6f^7\right)^2$ ______
11. $(-mnp)^5$ ______
12. $(0.2ab^2c)^2$ ______
13. $5a^2b(3ab^2 + 4a^3b^3 + 6a^4b)$ ______
14. $\frac{1}{2}ab^2c^3(4a^3b^2c - 18a^5b^3c^4 + 10a^2bc^4)$ ______
15. $-7x^4y^2(9xy^5 - 6x^3y^3 + 7x^5y)$ ______
16. $-\frac{2}{3}x^2y^2z^2(\frac{3}{4}xyz + \frac{9}{14}y^3z^5 - \frac{15}{22}x^4y^7)$ ______
17. $m^4n^5(-2m^2n^3 + 3m^3n^4 - 7mn^2)$ ______
18. $1.2ab^3(0.6b - 0.3a^2b^2)$ ______
19. $-3.4x^2y^3(0.7x^3 + 0.9y^2)$ ______
20. $-2.1m^3n(0.4n^3 - 0.21m^3)$ ______

Fill in the missing factor.

21. $(9a^2b^3c)$ ______ $= 72a^6b^6c^6$
22. ______ $(-11a^3bc^2) = 55a^6b^7c^8$
23. ______ $(3k^2 + 7b^2) = 9a^2k^3 + 21a^2kb^2$
24. $18r^3b^2($ ______ $+ 3r^6) = 72r^3b^4 + 54r^9b^2$

Solve.

25. If one pound of beans cost $35y$ cents, what is the cost of $8x$ pounds of beans? ______
26. If a car travels at the rate of $5x^2$ miles per hour, what is the distance traveled by the car in $7x$ hours? ______
27. A rectangle has its length equal to $5a^4b$ cm and width equal to $7ab$ cm. Find its area. ______

*Use with Lesson 15-4, text page 421.

Multiplication of Polynomials*

Name ____________________

Date ____________________

To multiply polynomials:

- Distribute each term of the first polynomial across each term of the second.
- Simplify the products by combining any similar terms.

$(x - 7)(x + 8) = \underline{?}$

$= (x \cdot x) + (x \cdot 8) + (-7)(x) + (-7)(8)$

$= x^2 + 8x - 7x - 56$

$= x^2 + 1x - 56$

$= x^2 + x - 56$

Multiply.

1. $(x + 3)(x + 4)$ ____________________

2. $(y + 11)(y + 2)$ ____________________

3. $(x - 7)(x - 6)$ ____________________

4. $(y - 10)(y - 8)$ ____________________

5. $(a + 13)(a - 5)$ ____________________

6. $(b - 9)(b + 8)$ ____________________

7. $(c + 12)(c - 7)$ ____________________

8. $(d - 15)(d + 3)$ ____________________

9. $(m + 8)(m + 5)$ ____________________

10. $(n + 14)(n + 7)$ ____________________

11. $(a - 10)(a + 10)$ ____________________

12. $(b + 16)(b - 16)$ ____________________

13. $(c + 12)(c + 12)$ ____________________

14. $(a + 13)(a + 13)$ ____________________

15. $(x - 11)^2$ ____________________

16. $(y - 17)^2$ ____________________

17. $(3x + 5)(2x + 3)$ ____________________

18. $(7x + 2)(6x - 5)$ ____________________

19. $(2x - 9)(3x - 4)$ ____________________

20. $(2x + 11)(x - 9)$ ____________________

21. $(8x - 3)(3x + 7)$ ____________________

22. $(9x + 2)(8x - 5)$ ____________________

23. $(4x + 3)(x - 6)$ ____________________

24. $(5x - 2)(7x - 5)$ ____________________

25. $(6x + 11)(6x - 11)$ ____________________

26. $(11x + 3)(11x - 3)$ ____________________

27. $(7x + 5)(7x + 5)$ ____________________

28. $(8x + 7)(8x + 7)$ ____________________

29. $(10x - 1)^2$ ____________________

30. $(15x - 2)^2$ ____________________

31. $(7a - 2b)(11a + 3b)$ ____________________

32. $(4a + 13b)(2a - b)$ ____________________

33. $(13a - 5b)(13a + 5b)$ ____________________

***Use with Lesson 15-5, text pages 422–423.**

Multiplication of Polynomials* (con't)

Name ______________________

Date ______________________

Multiply. Simplify the expressions.

34. $(m + 2n)(11m - 3n)$ ______

35. $(8a + 15b)(6a - 7b)$ ______

36. $(5a + 7b)^2$ ______

37. $(m + 3n)(m - 3n)$ ______

38. $(6c + d)(6c - d)$ ______

39. $(11a - 6b)^2$ ______

40. $(x + 2)(x^2 - 2x + 4)$ ______

41. $(c - 14d)^2$ ______

42. $(y + 4)(y^2 - 4y + 16)$ ______

43. $(a - 5)(a^2 + 5a + 25)$ ______

44. $(b - 6)(b^2 + 6b + 36)$ ______

45. $(x + 2)(x^2 + 4x + 4)$ ______

46. $(y - 4)(y^2 - 8y + 16)$ ______

47. $(a + 6)(a^2 - 3a + 1)$ ______

48. $(a - 7)(a^2 + 2a - 3)$ ______

49. $(a + 1)(6 - 5a - a^2)$ ______

50. $(3 - b)(b^2 - 6b + 5)$ ______

51. $(x + 3)(x + 5) - x^2$ ______

52. $9x^2 - (2x + 5)(3x - 4)$ ______

53. $x[(x + 4)(x - 4) + 5]$ ______

54. $(a + 5)^2 - (a - 2)^2$ ______

Solve.

55. Find the area of a rectangle if its length is $(3x - 7)$ in. and its width is $(x + 9)$ in. ______

56. A car travels at the rate of $(2x - 5)$ miles per hour. How far can it travel in $(6x - 7)$ hours? ______

57. Find the area of a square if the length of a side is $(11x - 2)$ cm. ______

58. Find the area of a circle in terms of π if its radius is $(5x - 4)$ cm. ______

*Use with Lesson 15-5, text pages 422–423.

Monomial Factors of Polynomials*

Name ____________________

Date ____________________

Find the GCF of $6x^2y + 45x^3y^2$.

$6x^2y = 2 \cdot 3 \cdot x \cdot x \cdot y$

$45x^3y^2 = 3 \cdot 3 \cdot 5 \cdot x \cdot x \cdot x \cdot y \cdot y$

So, GCF $= 3 \cdot x \cdot x \cdot y$ or $3x^2y$

$6x^2y + 45x^3y^2 = \underline{\ ?\ }$

GCF $= 3x^2y$

$\frac{6x^2y}{3x^2y} = 2 \qquad \frac{45x^3y^2}{3x^2y} = 15xy$

$6x^2y + 45x^3y^2 = 3x^2y(2 + 15xy)$

Factor.

1. $6x + 9y$ ____________
2. $21a - 28b$ ____________
3. $a^3b^3c^3 - a^2b^3c^4 + a^3b^2c^4$ ____________
4. $3m^3 + 4m^2n$ ____________
5. $5r^6 + 10a^2$ ____________
6. $x^5y^3z + x^3y^4z^2 + x^3y^2z$ ____________
7. $25a^6 + 10a^2$ ____________
8. $18x^4 - 21x^3$ ____________
9. $125x^5 + 175x^4 + 200x^2$ ____________
10. $a^3b^2 + abc$ ____________
11. $p^3q^4 - p^2q^3r$ ____________
12. $120a^5 - 140a^4 + 180a^3$ ____________
13. $36x^3y^5z^4 + 48x^5y^2z^2$ ____________
14. $49a^7b^4c^3 - 84a^5b^4c^5$ ____________
15. $p^8 + p^{12} + p^{20}$ ____________
16. $11m^4n^5 + 121m^3n^2$ ____________
17. $13a^5b^5 - 169a^4b^3$ ____________
18. $x^{19} + x^{27} + x^{13}$ ____________
19. $147 - 36x^5 + 288x^7$ ____________
20. $76x^3y^5 - 95x^2y^4z^2 + 133x^4y^3z$ ____________
21. $189x^9 + 270x^4 + 144$ ____________
22. $a^5b^3c^3 + a^4b^3c^4 + a^5b^2c^4$ ____________
23. $26a^3b^2 + 39a^2bc + 65ab^2c^2$ ____________
24. $p^6q^3r^4 - p^5q^3r^5 + p^6q^2r^5$ ____________
25. $32x^5y^3 + 40x^3y^4 + 48x^2y^5$ ____________
26. $49xyz - 56x^2z^3 + 147xy^2z$ ____________

***Use with Lesson 15-6, text pages 424–425.**

Monomial Factors of Polynomials (con't)

Name ______________________
Date ______________________

Factor.

27. $40c^4d^2 - 50c^2d^3 + 10cd^3$ ______________________

28. $1.5m^3n - 4.5mn^2 + 6mn$ ______________________

29. $-10x^5y - 75x^4y^2 - 45x^3y^3$ ______________________

30. $1.3c^3d^2 + 2.6cd^3 - 6.5cd^2$ ______________________

31. $-33a^5b^3c^2 - 44a^4b^2c^3 - 22a^5bc^4$ ______________________

32. $90m^5n^5c^6 - 60m^2n^4c^3 - 50m^5n^4c^5$ ______________________

33. $0.2x^3y^5z^2 - 0.8x^2y^4z^3 + 1.2x^4y^2z^4$ ______________________

34. $120x^3y^5z - 75x^2y^6z^2 + 135x^3y^3z^4$ ______________________

35. $2.7a^4b^3c + 0.3a^3b^3c^2 - 0.6a^5b^2c^2$ ______________________

36. $119p^6q^8r^2 + 68p^8q^4r^4 + 170p^5q^4r^6$ ______________________

Solve.

37. The area of a parallelogram is represented by $(2x^2y - 2xy^2)$ in.^{2}. Express the area as a product of two factors. ______________________

38. The area of a rectangle is $(8a^2bc - 4abc^2)$ cm^2. Express the area as a product of two factors. ______________________

39. Jessica uses $4c^4 + 96$ to represent the area of a rectangle and 4 to represent its length. What expression represents its width? ______________________

40. Charlene uses $48b^2 - 32b^2c^2$ to represent the area of a parallelogram. If the GCF of $48b^2 - 32b^2c^2$ is the base, what is the base and the height of the parallelogram? ______________________

***Use with Lesson 15-6, text pages 424–425.**

Factoring Trinomials*

Name ____________________

Date ____________________

$x^2 - x - 6 = \underline{\ ?\ }$
Factors of 6: {1, 2, 3, 6}
Factors of −6 whose sum is −1: 2 and −3
$x^2 - x - 6 = (x + 2)(x - 3)$

Factor.

1. $x^2 + 6x + 5$
2. $a^2 + 12a + 27$
3. $y^2 - 20y + 19$
4. $x^2 + 7x + 12$
5. $b^2 - 20b + 64$
6. $y^2 - 38y - 39$
7. $y^2 - 11y + 24$
8. $b^2 + 11b - 102$
9. $a^2 + 17a + 16$
10. $y^2 - 10y + 21$
11. $x^2 - 15x - 54$
12. $a^2 - 24a + 44$
13. $a^2 - 3a - 10$
14. $x^2 - 19x - 42$
15. $x^2 - 41x + 180$
16. $a^2 + 3a - 28$
17. $y^2 + 22y + 120$
18. $x^2 - 7x - 144$
19. $m^2 - 5m - 36$
20. $y^2 + 26y + 153$
21. $c^2 - 15cd + 44d^2$
22. $n^2 + 9n - 52$
23. $y^2 - 26y + 144$
24. $x^4 - 16x^2y^2 + 55y^4$
25. $n^2 - 10n + 9$
26. $m^2 - 2m - 143$
27. $m^4 + 8m^2n^2 - 105n^4$
28. $c^2 - 12c + 11$
29. $x^2 + 23x + 120$
30. $a^4 - 19a^2b^2 + 34b^4$
31. $d^2 + 18d + 65$
32. $n^2 - 13n + 42$
33. $c^4 - 22c^2d^2 - 23d^4$
34. $r^2 - 19r + 70$
35. $x^2 - 17x + 60$
36. $p^4 + 14p^2q^2 - 32q^4$

*Use with Lesson 15-7, text pages 426–427.

Factoring Trinomials* (con't)

Name ______________________

Date ______________________

Factor.

37. $a^2 + 6ab + 8b^2$

38. $m^4 - 9m^2n^2 - 70n^4$

39. $x^2 - 12xy + 32y^2$

40. $x^2y^2 + 17xy + 66$

41. $b^2 - 8bc - 20c^2$

42. $x^2y^2 + 15xy - 34$

43. $y^2 + 13yz - 48z^2$

44. $a^2b^2 + 19ab + 70$

45. $m^2 + 9mn + 14n^2$

46. $c^2d^2 - 22cd - 75$

47. $r^2 - 15rs + 50s^2$

48. $x^4y^4 - 3x^2y^2 - 108$

49. $a^2 + 16ab + 63b^2$

50. $a^4b^4 - 11a^2b^2 - 60$

Solve.

51. The area of a rectangle is $(x^2 - 7x + 12)$ cm^2. Find the binomials that represent the dimensions of the rectangle.

52. The distance traveled by a bus is $(x^2 - 5x - 24)$ mi. Find the binomials that represent the rate of the bus per hour, and the time in hours the bus travels to cover the given distance.

53. The area of a parallelogram is $(x^2 + 17x + 42)$ $in.^2$. Find the binomials that represent the dimensions of the parallelogram.

*Use with Lesson 15-7, text pages 426–427.

Factoring More Polynomials*

Name ______________________

Date ______________________

$10x^2 + 3x - 7 = \underline{\ ?\ }$

Factors of 10: {1, 2, 5, 10} and factors of 7: {1, 7}
Work out possible combinations of factors to find $3x$ as the middle term.

$10x^2 \boxed{- 7x + 10x} - 7 = (x + 1)(10x - 7)$

Factor.

1. $3x^2 + 4x + 1$
2. $4x^2 - x - 3$
3. $60 - 11x - 14x^2$
4. $14x^2 - 9x + 1$
5. $6x^2 - 7x - 10$
6. $72 + 13x - 15x^2$
7. $15a^2 - 11a + 2$
8. $2a^2 + a - 10$
9. $24r^2 + 38rs + 15s^2$
10. $12a^2 - a - 6$
11. $30a^2 - 17a + 2$
12. $35x^2 - 48xy - 27y^2$
13. $35x^2 + 31x + 6$
14. $4x^2 + x - 14$
15. $21a^2 - 58ab - 40b^2$
16. $10x^2 + 13x - 3$
17. $9x^2 - 9x - 4$
18. $15m^2 - 106mn + 7n^2$
19. $16x^2 + 6x - 1$
20. $22x^2 - 17x + 3$
21. $24c^2 - 19cd - 9d^2$
22. $8x^2 + 22x + 15$
23. $20x^2 - 32x + 3$
24. $15x^2 - 34xy + 15y^2$
25. $5x^2 + 4x - 1$
26. $11x^2 - 9x - 2$
27. $45a^2 + 56ab - 45b^2$
28. $5x^2 + 21x + 4$
29. $12x^2 - 20x + 7$
30. $24x^4 + 26x^2y + 5y^2$
31. $15x^2 - 26x + 8$
32. $6x^2 - 13x - 63$
33. $6x^6 + 7x^3y - 55y^2$

*Use with Lesson 15-8, text pages 428–429.

Factoring Polynomials* (con't)

Name ______________________

Date ______________________

34. $132r^2 - 101rs + 14s^2$

35. $77x^2 - 170xy + 77y^2$

36. $36s^2 + 23st - 3t^2$

37. $6a^2 + 5ab - 6b^2$

38. $39a^2 + 19ab + 2b^2$

39. $15x^4 - 22x^2y + 8y^2$

40. $84m^2 + 16mn - 5n^2$

41. $19x^6 - 59x^3y + 6y^2$

42. $2c^2 + 13cd - 45d^2$

43. $12x^4 - 40x^2y^2 + 25y^4$

44. $8x^4 + 26x^2y^2 + 21y^4$

45. $6x^6 + 7x^3y^3 - 10y^6$

46. $6x^4 + 43x^2y^2 + 77y^4$

47. $18h^4 + 23h^2b^2 - 6b^4$

48. $15x^6 + 41x^3y^3 + 14y^6$

49. $35a^6 - 33a^3b^3 - 8b^6$

Solve.

50. The area of a rectangle is $(6x^2 - 7x - 5)$ cm^2. Find the binomials that represent the dimensions of the rectangle.

51. The area of a rhombus is $(2x^2 + 13x + 21)$ in.^{2}. Find the binomials that represent the dimensions of the rhombus.

*Use with Lesson 15-8, text pages 428–429.

Perfect Square Trinomials*

Name ______________________

Date ______________________

Simplify.

1. $(x + 5)^2$ ______________

2. $(4c + 5d)^2$ ______________

3. $(3a - 2)^2$ ______________

4. $(y - 8)^2$ ______________

5. $(m + 2n)^2$ ______________

6. $(7c - 3d)^2$ ______________

7. $(2a + 3)^2$ ______________

8. $(x + 7)^2$ ______________

9. $(m - 9n)^2$ ______________

$4x^2 - 20x + 25 = \underline{\ ?\ }$

- 1st term: $4x^2 \longrightarrow$ perfect square
- 3rd term: $25 \longrightarrow$ perfect square
- middle term: $20x \longrightarrow 20x = 2 \cdot \sqrt{4x^2} \cdot \sqrt{25} = 2 \cdot 2x \cdot 5$

Factors are: $(2x - 5)(2x - 5) = (2x - 5)^2$

Factor.

10. $x^2 + 22x + 121$ ______________

11. $x^2 + 14x + 49$ ______________

12. $y^2 - 24y + 144$ ______________

13. $y^2 - 20y + 100$ ______________

14. $9x^2 - 30x + 25$ ______________

15. $4x^2 - 28x + 49$ ______________

16. $16x^2 + 40xy + 25y^2$ ______________

17. $9a^2 - 66ab + 121b^2$ ______________

18. $a^2b^2 - 10ab + 25$ ______________

19. $0.04x^2 + 1.2xy + 0.09y^2$ ______________

20. $1.21x^2 - 1.1xy + 0.25y^2$ ______________

Solve.

21. The area of a square is $(121x^2 + 22x + 1)$ cm^2. Find the binomial that represents the length of a side of the square. ______________

22. The area of a circle is $(9x^2 + 24x + 16)\ \pi$ cm^2. Find the binomial that represents the radius of the circle. ______________

*Use with Lesson 15-9, text page 430.

Difference of Two Squares*

Name ______________________

Date ______________________

$(x + 4)(x - 4) = \underline{\ ?\ }$
$(x + 4)(x - 4) = x^2 - 4x + 4x - 16$
$= x^2 - 16$

Multiply.

1. $(x - 5)(x + 5)$ ______
2. $(x + 13)(x - 13)$ ______
3. $(14a + 1)(14a - 1)$ ______
4. $(x + 4y)(x - 4y)$ ______
5. $(x - 7y)(x + 7y)$ ______
6. $(\frac{2}{3}a + \frac{1}{2}b)(\frac{2}{3}a - \frac{1}{2}b)$ ______
7. $(3a + 5b)(3a - 5b)$ ______
8. $(2a - 13b)(2a + 13b)$ ______
9. $(\frac{4}{7}c + \frac{2}{3}d)(\frac{4}{7}c - \frac{2}{3}d)$ ______

$4x^2 - 25 = \underline{\ ?\ }$
$\sqrt{4x^2} = 2x$ and $\sqrt{25} = 5$
Factors are: $(2x + 5)(2x - 5)$

Factor.

10. $x^2 - 81$ ______
11. $x^2 - 16$ ______
12. $m^2 - n^2$ ______
13. $9 - 25a^2$ ______
14. $25 - 121b^2$ ______
15. $\frac{1}{4}x^2 - \frac{1}{9}y^2$ ______
16. $64x^2 - 9y^2$ ______
17. $169x^2 - 25y^2$ ______
18. $0.36x^2 - y^2$ ______
19. $a^2b^2 - c^2$ ______
20. $\frac{16}{25}m^2 - \frac{4}{9}n^2$ ______
21. $1.21c^2 - 0.09d^2$ ______

Solve.

22. The area of a rectangle is $(9x^2 - 16y^2)$ cm^2. Express the area as a product of two binomials. ______

23. The area of a rhombus is $(169x^2 - 4y^2)$ cm^2. Express the area as a product of two binomials. ______

24. The average mark of a class is represented by the expression $(144a^2 - 225b^2)$. Express the average mark as a product of two binomials. ______

*Use with Lesson 15-10, text page 431.

Complete Factorization*

Name ______________________

Date ______________________

To factor a polynomial completely:

- Factor out any GCF.
- Divide each term by the GCF.
- Factor, if possible, the new polynomial.
- Rewrite the original polynomial as a product of the GCF and these factors.

Factor completely.

1. $14x + 35y$

2. $18x + 27y$

3. $15ab + 35bc$

4. $20mn + 50np$

5. $5x^2 - 20y^2$

6. $3x^2y - 27y$

7. $m^3 - mn^4$

8. $x^3y^5 - xy$

9. $4x^2y - 8xy + 4y$

10. $2a^2x^2 - 8a^2x + 8a^2$

11. $3x^3 - 3x^2 - 6x$

12. $2x^2y - 10xy + 12y$

13. $5ax^2y^2 - 5az^2$

14. $28a^3x^2y^2 - 7a^3c^2$

15. $5m^2 - 30mn + 45n^2$

16. $7a^2 - 70ab + 175b^2$

17. $2a^3 - a^2 - a$

18. $15b^2 - 5b - 10$

19. $18b + 21b^2 + 6b^3$

20. $50ab - 5ab^2 - 15ab^3$

21. $24x^3y - 28x^2y^2 + 8xy^3$

22. $10x^3y^2 + 34x^2y^3 + 12xy^4$

*Use with Lesson 15-11, text pages 432–433.

Complete Factorization* (con't)

Name ______________________

Date ______________________

Factor completely.

23. $3y^2 + 6$

24. $r + 2sr$

25. $mx^2 - mx - 2m$

26. $8x + 24$

27. $mn^2 - m^2n$

28. $15a^2 + 10a - 5$

29. $x^2 + x$

30. $6b^2 - 30b$

31. $2\pi r - \pi r^2$

Factor completely. If the polynomial cannot be factored, label it prime.
(Hint: $5c^2 + 8e^2$ is prime.)

32. $98x^3y - 28x^3y^2 + 2x^3y^3$

33. $243x^2y^2 - 54x^2y^3 + 3x^2y^4$

34. $a^4 - b^4$

35. $36x^2y^2 + 84xyz + 49z^2$

36. $81x^4 - y^4$

37. $81a^2c^2 + 72abc + 64b^2$

38. $36a^4 - 13a^2 + 1$

39. $4m^4 + 144n^6$

40. $100a^4 - 29a^2 + 1$

41. $36a^4 - 225b^4$

42. $6a^4b^4 - 17a^2b^2 + 7$

43. $44x^2y^2 - 3xyz - 65z^2$

44. $72c^4d^4 + 35c^2d^2 + 3$

45. $42a^2b^2 + 25abc - 28c^2$

46. $6x^4y^4 - x^2y^2 - 77$

47. $6r^2p^2 + rpt + 40t^2$

48. $121a^2b^2 - 264abc + 144c^2$

49. $6a^3r^2s^2 - 54a^3t^4$

50. $90c^2d^2 + 129cde + 28e^2$

51. $6n^6y + 216y$

*Use with Lesson 15-11, text pages 432–433.

Complete Factorization* (con't)

Name ______________________

Date ______________________

Factor. Rearrange terms if necessary.

52. $x^2 + 6 + 5x$

53. $-3y + y^2 - 4$

54. $a^2 - 5a - 6$

55. $3y + y^2 - 4$

56. $y^2 + 4 - 5y$

57. $x^2 - 9x + 14$

58. $-8c - 9 + c^2$

59. $x^2 + 5x - 14$

60. $9 + d^2 + 6d$

61. $x^2 - 5x - 14$

62. $10r + 24 + r^2$

63. $y^2 + 48y - 100$

64. $16 + 10m + m^2$

65. $x^2 - 12x + 20$

66. $x^4 + 2x^2 + 1$

67. $y^2 + 100 + 29y$

68. $x^6 + 1 + 2x^3$

69. $2 - y - y^2$

70. $6y^3 + 3y^2 - 3y$

71. $x^4 - 8x^2 - 9$

72. $100x^2 - 4y^2$

Factor completely.

73. $3a - a^2$

74. $x^3 - x^2 - 2x$

75. $2m^3 - 11$

76. $r^2 + 11r$

77. $9x^5 + 27x^3 - 45x^2$

78. $4x^2 - 36y^2$

79. $x^2y^2 + y$

80. $14a^2c - 7a^2$

81. $3x^4 + 11x^2 + 13x$

82. $16c^4 - 81$

83. $8n^2 + 2nr - r^2$

84. $(a + b)^2 - (c + d)^2$

85. $(e + f)^2 + 6(e + f) + 9$

86. $576a^2 - 1$

87. $9r^2 - 9m^2$

*Use with Lesson 15-11, text pages 432–433.

Complete Factorization* (con't)

Name ______________________

Date ______________________

Solve. Show your work.

88. The volume of a rectangular prism is $(6x^3y - 6x^2y - 120xy)$ cm^3. Write in factored form the expression that represents the dimensions of the rectangular solid. ______________

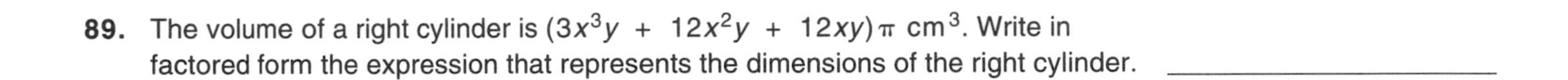

89. The volume of a right cylinder is $(3x^3y + 12x^2y + 12xy)\pi$ cm^3. Write in factored form the expression that represents the dimensions of the right cylinder. ______________

90. If the product of three factors is $(6x^3 + 12x^2 - 48x)$, what are the factors? ______________

91. What are the factors representing three consecutive numbers if the product is $x^3 + 3x^2 + 2x$? ______________

92. What factor, when squared, equals one half of $(2x^2 + 4x + 2)$? ______________

*Use with Lesson 15-11, text pages 432–433.

Division of Polynomials*

Name ______________________

Date ______________________

$$\frac{a^7}{a^5} = \underline{?} = a^{7-5} = a^2$$

Divide.

1. $\frac{a^{10}}{a^6}$ ______

2. $\frac{x^{12}y^6}{x^4y^2}$ ______

3. $\frac{6x^3y}{2xy}$ ______

4. $\frac{39a^2bc}{13abc}$ ______

5. $\frac{-24x^3y}{6x^2y}$ ______

6. $\frac{35a^4b^2}{-7a^3b}$ ______

7. $\frac{22.5m^5n^4}{1.5m^2n^3}$ ______

8. $\frac{1.69c^6d^4}{1.3c^3}$ ______

9. $\frac{-336c^5d^7}{-16c^2d^5}$ ______

10. $\frac{-495e^8f^3}{-45e^3f}$ ______

11. $\frac{a^3b^3 + a^2b^2}{a^2b^2}$ ______

12. $\frac{27c^3d^5 - 18c^2d^6}{9c^2d^3}$ ______

13. $\frac{-136x^2y^3 + 187xy}{17xy}$ ______

14. $\frac{x^4y^5 + x^3y^2}{x^2y^2}$ ______

15. $\frac{-196a^7 + 126a^6}{14a^5}$ ______

16. $\frac{3\pi r^2h - 9\pi rh}{-3\pi rh}$ ______

17. $\frac{-75x^{10} + 90x^{12}}{-15x^9}$ ______

18. $\frac{x^3y^4z^2 + x^5y^3z^3 - x^4y^5z^4}{x^3y^3z^2}$ ______

19. $\frac{-209r^3s^2 + 95r^5s^2}{19rs^2}$ ______

***Use with Lesson 15-12, text pages 434–435.**

Division of Polynomials* (con't)

Name ______________________

Date ______________________

$(x^2 - x - 20) \div (x + 4) = \underline{\ ?\ }$

$$\begin{array}{r} x \quad -5 \\ x + 4 \overline{)\, x^2 \quad -x \quad -20} \\ \underline{\mp x^2 \ \mp 4x} \\ -5x \ -20 \\ \underline{\pm 5x \ \pm 20} \end{array}$$

Check:

$(x - 5)(x + 4) = x^2 - x - 20$

Divide and check.

20. $(a^2 + 5a + 6) \div (a + 3)$ ______________________

21. $(10 - 11b + 3b^2) \div (5 - 3b)$ ______________________

22. $(r^2 + 7r + 12) \div (r + 4)$ ______________________

23. $(56 - 3m - 20m^2) \div (8 - 5m)$ ______________________

24. $(2x^2 - 13x - 7) \div (2x + 1)$ ______________________

25. $(m^3 + 9m^2 + 19m + 10) \div (m + 2)$ ______________________

26. $(6x^2 - 7x - 20) \div (3x + 4)$ ______________________

27. $(r^3 - 10r^2 + 25r + 12) \div (r - 3)$ ______________________

28. $(3x^2 - 11x - 30) \div (x - 6)$ ______________________

29. $(9x^3 - 9x^2y + 17xy^2 - 5y^3) \div (3x - y)$ ______________________

30. $(10x^2 - 19x + 12) \div (2x - 3)$ ______________________

31. $(4a^3 + 12a^2b + ab^2 - 12b^3) \div (2a + 3b)$ ______________________

Solve.

32. The area of a rectangle is represented by $(16r^2 + 34r - 15)$ cm^2. If the length of the rectangle is represented by $(8r - 3)$ cm, what is the width of the rectangle? ______________________

33. The distance traveled by a car is $(2x^2 - 7x + 6)$ miles. If the car travels $(2x - 3)$ miles per hour, how much time did the car travel to cover the distance? ______________________

*Use with Lesson 15-12, text pages 434–435.

Multiplication and Division of Rational Expressions*

Name ______________________

Date ______________________

Simplify $\frac{5x^2}{20x^3}$. $\quad \frac{5x^2}{20x^3} = \frac{(\cancel{5})(\cancel{x})(\cancel{x})}{(4)(\cancel{5})(\cancel{x})(\cancel{x})(x)} = \frac{1}{4x}$

Simplify.

1. $\frac{6ab}{24ac}$ ______

2. $\frac{^{-}15x^2y}{25x^3y^2}$ ______

3. $\frac{25 - x^2}{2x - 10}$ ______

4. $\frac{2a + 6}{4}$ ______

5. $\frac{3b + 9}{-15}$ ______

6. $\frac{b - a}{a - b}$ ______

7. $\frac{12xy - 3y^2}{3xy}$ ______

8. $\frac{14a^2b + 21ab^2}{28ab}$ ______

9. $\frac{x^2 - 5x - 6}{x^2 + 2x + 1}$ ______

10. $\frac{5m^2}{15m^2 - 5mn}$ ______

11. $\frac{9d}{9d + 18}$ ______

12. $\frac{3y^2 - 10y + 3}{9y^2 - 6y + 1}$ ______

$$\frac{5}{2x - 12} \cdot \frac{x - 6}{x^2 - 10} = \frac{5}{2\cancel{(x - 6)}} \cdot \frac{\cancel{(x - 6)}}{(x^2 - 10)} = \frac{5}{2(x^2 - 10)} = \frac{5}{2x^2 - 20}$$

Multiply

13. $\frac{7x^2}{9y^2} \cdot \frac{18xy^2}{7x^3}$ ______

14. $\frac{36a^2}{15b} \cdot \frac{3b}{6a}$ ______

15. $\frac{7}{8} \cdot \frac{2a + 4}{21}$ ______

16. $\frac{4a + 12}{16a} \cdot \frac{a^3}{24}$ ______

17. $\frac{a^2 - 1}{a^2} \cdot \frac{15}{3a^2 - 3a}$ ______

18. $\frac{x^2 - 9}{3} \cdot \frac{12}{2x^2 - 6}$ ______

*Use with Lesson 15-13, text pages 436–437.

Multiplication and Division of Rational Expressions* (con't)

Name ______________________

Date ______________________

Multiply.

19. $\frac{5x + 10}{15x + 30} \cdot \frac{6x + 12}{x^2 - 4}$

20. $\frac{y^2 - 3y + 2}{2y^3} \cdot \frac{4y^2}{2y - 4}$

21. $\frac{x^2 + 2x - 15}{x^2 + 9x + 20} \cdot \frac{x^2 + 3x - 4}{x^2 - 4x + 3}$

22. $\frac{6x^2 - 7x - 3}{10x^2 - 17x + 3} \cdot \frac{3x^2 + 16x + 16}{3x^2 + 13x + 4}$

$$\frac{5x}{8} \div \frac{6x}{x + 2} = \frac{5\not{x}}{8} \cdot \frac{x + 2}{6\not{x}} = \frac{5x + 10}{48}$$

Divide.

23. $\frac{4x}{7y} \div \frac{10x^3}{28y^2}$

24. $\frac{7xy^2}{10ab} \div \frac{14y^3}{5a^2b^2}$

25. $\frac{5x^2y^2}{15z} \div 10xy$

26. $\frac{14x^3y^3}{7z^3} \div 28x^2y^2$

27. $\frac{a^2 - 1}{7} \div \frac{a + 1}{14}$

28. $\frac{a^2 - 2a + 1}{6a} \div \frac{a - 1}{15a^2}$

29. $\frac{a^2 - 2a - 15}{3a} \div \frac{a^2 - 9}{a^2}$

30. $\frac{2b^2 - b - 1}{b^2 - 2b + 1} \div \frac{4b^2 + 4b + 1}{2b^2 + 7b + 3}$

31. $\frac{x^2 - 6x + 9}{4x - 12} \div (3 - x)$

32. $(16 - y^2) \div \frac{y^2 + 9y + 20}{2y + 10}$

Simplify.

33. $\frac{2x + 6}{x^2 - 9} \div \frac{x + 3}{3 - x} \cdot \frac{x + 3}{4}$

34. $\frac{x^2 + 2xy + y^2}{x^2 - y^2} \div \frac{x + y}{y^2 - x^2} \cdot \frac{x - y}{x^2 - 2xy + y^2}$

35. For what value(s) of x is $\frac{x^2 - 49}{7} \cdot \frac{14}{2x - 14}$ undefined? ______________

36. For what value(s) of y is $\frac{y^2 - 4y + 4}{y^2} \div \frac{y^2 - 4}{y}$ undefined? ______________

*Use with Lesson 15-13, text pages 436–437.

Addition and Subtraction with Like Denominators*

Name ______________________

Date ______________________

To add or subtract rational expressions with *like* denominators:

- Add or subtract the numerators.
- Write the sum or difference over the like denominator.
- Simplify.

Add or subtract. Simplify answers where possible.

1. $\frac{9}{4x} + \frac{5}{4x}$ ______________________

2. $\frac{2a}{5x} + \frac{8a}{5x}$ ______________________

3. $\frac{11x}{x - y} - \frac{11y}{x - y}$ ______________________

4. $\frac{6x}{x - 3} - \frac{18}{x - 3}$ ______________________

5. $\frac{8}{5y} + \frac{4}{5y} - \frac{7}{5y}$ ______________________

6. $\frac{6}{9c} + \frac{7}{9c} - \frac{4}{9c}$ ______________________

7. $\frac{x^2 + 4x}{x^2 - 4} + \frac{4}{x^2 - 4}$ ______________________

8. $\frac{6x - 5}{x^2 - 1} - \frac{5x - 6}{x^2 - 1}$ ______________________

9. $\frac{x^2 + 3xy}{x + y} + \frac{y^2 - xy}{x + y}$ ______________________

10. $\frac{a^2 - 2ab}{a - 2b} - \frac{ab - 2b^2}{a - 2b}$ ______________________

11. $\frac{x + 4y}{x^2 - y^2} + \frac{4x - 7y}{x^2 - y^2} - \frac{3x - y}{x^2 - y^2}$ ______________________

12. $\frac{7 + 3x}{x^2 - 4} - \frac{8 + 4x}{x^2 - 4} + \frac{3 + 2x}{x^2 - 4}$ ______________________

13. $\frac{3a^2 - 2b^2}{a^2 + y^2} - \frac{2a^2 - 3y^2}{a^2 + y^2}$ ______________________

14. $\frac{5c - 2}{3c - 4} - \frac{16c - 26}{3c - 4}$ ______________________

*Use with Lesson 15-14, text page 438.

Addition and Subtraction with Like Denominators* (con't)

Name ____________________

Date ____________________

Add or subtract. Simplify answers.

15. $\frac{a^2 + 4a}{a^2 - a - 6} + \frac{8 - a^2}{a^2 - a - 6}$

____

16. $\frac{4b^2 + 7b}{2b^2 + 5b + 2} - \frac{1 + 7b}{2b^2 + 5b + 2}$

____

17. $\frac{7x^2 + 12x}{4x^2 + 3x} - \frac{5x^2 + 3x}{4x^2 + 3x}$

____

18. $\frac{3r + 1}{r^2 - 7r + 12} + \frac{r + 17}{r^2 - 7r + 12}$

____

Solve. Show your work.

19. What is the average rate of speed for a 6-hour trip where the first 160 miles are by car and the next 110 miles are by bus?

____

20. Find the perimeter of a triangle whose sides are represented by $\frac{4x + 1}{x + 6}$, $\frac{2 - 3x}{x + 6}$, and $\frac{9 + x}{x + 6}$.

____

21. The perimeter of a triangle is represented by $\frac{6a - 7}{6}$, and the two sides are represented by $\frac{2a - 6}{6}$ and $\frac{a + 1}{6}$. Find the representation of the third side.

____

*Use with Lesson 15-14, text page 438.

Least Common Denominator of Rational Expressions*

Name ______________________

Date ______________________

To find the Least Common Denominator (LCD) of two or more rational expressions:

- Completely factor each denominator.
- Express every prime factor with the greatest exponent it has in any denominator.
- Multiply these prime factors with their greatest exponents. This product is the LCD of the given expressions.

Find the LCD.

1. $\frac{1}{x}, \frac{3}{4x}$ ______________________

2. $\frac{2}{15a}, \frac{4}{30a^2}$ ______________________

3. $\frac{3}{7x^2y}, \frac{5}{14xy^2}$ ______________________

4. $\frac{9}{3x + 2}, \frac{7}{3x^2 + 5x + 2}$ ______________________

5. $\frac{14}{2x - 7}, \frac{8x}{4x^2 - 28x + 49}$ ______________________

6. $\frac{4}{5x + 10z}, \frac{7}{3x + 6z}$ ______________________

7. $\frac{4x}{x - 3y}, \frac{9}{3y - x}$ ______________________

8. $\frac{5a}{a + 1}, \frac{2a}{a + 5}$ ______________________

9. $\frac{8}{x - 5}, \frac{4}{x}$ ______________________

10. $\frac{7}{x^2 - xy}, \frac{10}{xy - y^2}$ ______________________

Find the LCD and complete the fraction.

11. $\frac{4}{x - 4} = \frac{\quad}{x^2 - 16}$

12. $\frac{2}{3xy} = \frac{\quad}{9x^3y^2}$

13. $\frac{2a + b}{a - b} = \frac{\quad}{ab - b^2}$

14. $\frac{7}{r^3} = \frac{\quad}{t^2r^3}$

15. $\frac{3}{x - 5} = \frac{\quad}{x^2 - 7x + 10}$

16. $\frac{5}{2a - 4} = \frac{\quad}{2a^2 - 8}$

*Use with Lesson 15-15, text page 439.

Addition and Subtraction with Unlike Denominators*

Name ______________________

Date ______________________

To add and subtract rational expressions with *unlike* denominators:

- Find the LCD.
- Change each rational expression to an equivalent fraction with the LCD as the denominator.
- Add or subtract.
- Simplify, if necessary.

Add or subtract.

1. $\frac{9x}{8y} - \frac{3x}{4y}$ ______________________

2. $\frac{2}{5x} + \frac{3}{10x}$ ______________________

3. $\frac{4}{x^2} - \frac{7}{x}$ ______________________

4. $\frac{1}{xy} + \frac{1}{yz}$ ______________________

5. $\frac{7}{ab} - \frac{9}{ac}$ ______________________

6. $\frac{a}{3xy} - \frac{b}{2yz}$ ______________________

7. $\frac{a + 5}{2a} + \frac{2a - 1}{4a}$ ______________________

8. $\frac{m + 6}{m} + \frac{m - 3}{4m}$ ______________________

9. $\frac{3d - 7}{2d} - \frac{3d - 2}{3d^2}$ ______________________

10. $\frac{9}{xy} + \frac{2}{yz} - \frac{3}{xz}$ ______________________

11. $x + \frac{1}{x}$ ______________________

12. $x + 1 + \frac{1}{x + 1}$ ______________________

13. $x - 4 - \frac{x}{x + 2}$ ______________________

14. $x - \frac{1}{x - 3}$ ______________________

15. $\frac{6}{3x - 5y} + \frac{9}{5y - 3x}$ ______________________

16. $\frac{4}{x - 4} + \frac{8}{2x - 8}$ ______________________

17. $\frac{5}{4a - 3} + \frac{8}{8a - 6}$ ______________________

18. $\frac{1}{x - 3} + \frac{1}{x + 3}$ ______________________

*Use with Lesson 15-16, text pages 440–441.

Addition and subtraction with Unlike Denominators* (con't)

Name ______________________

Date ______________________

Add or subtract.

19. $\dfrac{9}{y + 5} - \dfrac{6}{y - 5}$ ______________________

20. $\dfrac{x^2}{x^2 - y^2} - \dfrac{xy}{y - x}$ ______________________

21. $\dfrac{7}{x^2 - xy} + \dfrac{5}{xy - y^2}$ ______________________

22. $\dfrac{7a}{a^2 + 2a - 3} - \dfrac{3a + 4}{a^2 + 5a + 6}$ ______________________

23. $\dfrac{4}{x^2 - 9} - \dfrac{7}{x^2 - 5x + 6}$ ______________________

24. $\dfrac{3x}{x^2 - 4} + \dfrac{4x - 3}{x^2 + 2x - 8}$ ______________________

25. $\dfrac{1}{x^2 - x - 2} + \dfrac{2}{3x^2 + 2x - 1} - \dfrac{1}{3x^2 - 7x + 2}$ ______________________

26. $\dfrac{2x}{4x^2 + x - 3} - \dfrac{1}{x^2 + 6x + 5} + \dfrac{x + 1}{4x^2 + 17x - 15}$ ______________________

Solve.

27. Find the perimeter of a triangle if the lengths of its sides are represented by $\dfrac{1}{x - 2}$, $\dfrac{1}{x - 3}$, and $\dfrac{1}{x + 5}$. ______________________

28. The perimeter of a rectangle is represented by $\dfrac{4x^2 + 4}{x^2 - 4x - 21}$ and its length by $\dfrac{x + 1}{x + 3}$. Find its width. ______________________

***Use with Lesson 15-16, text pages 440–441.**

Linear Equations and Slope*

Name ______________________

Date ______________________

Remember: To find the slope of a line represented by a linear equation:

- Choose any 2 points, P_1 and P_2, on the line. Name their coordinates, $P_1(x_1, y_1)$ and $P_2(x_2, y_2)$.
- Substitute the value of these coordinates in the slope formula:

$$m = \frac{(y_2 - y_1)}{(x_2 - x_1)}$$

Find the slope of each line (1–4) in the given graph.

1. ______________

2. ______________

3. ______________

4. ______________

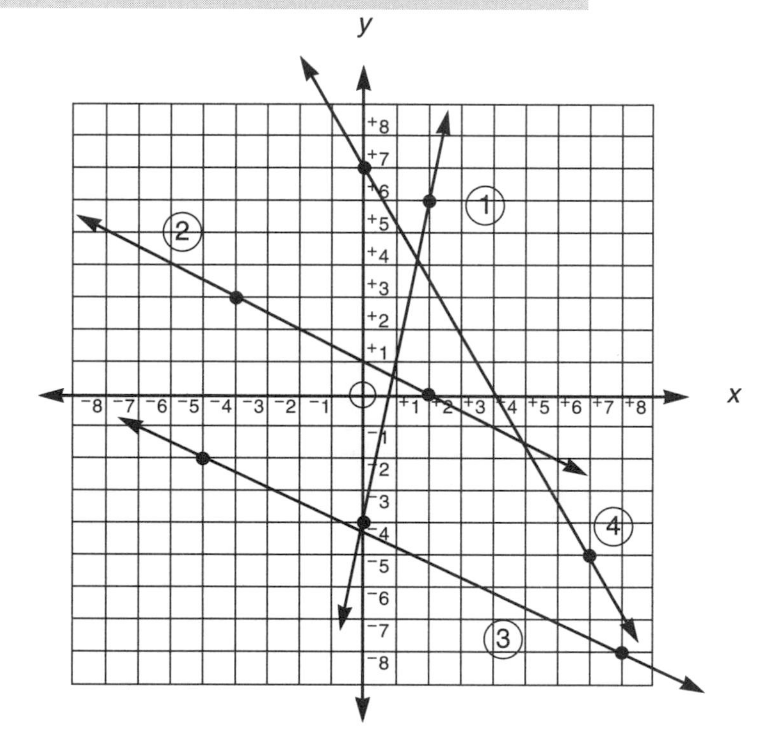

Find the slope of the line that passes through each pair of points.

5. $(2, 5)$ and $(6, 10)$ ______________

6. $(^{-}3, 7)$ and $(4, ^{-}5)$ ______________

7. $(^{-}1, ^{-}8)$ and $(^{-}9, ^{-}2)$ ______________

8. $(0, 7)$ and $(14, 0)$ ______________

9. $(0, 0)$ and $(8, 16)$ ______________

10. $(1, ^{-}6)$ and $(^{-}8, ^{-}9)$ ______________

*Use with Lesson 16-1, text pages 456–457.

Linear Equations and Slope* (cont'd)

Name ______________________

Date ______________________

Find the slope of each line.

11. $2x + 3y = 5$ ______________________

12. $5x - y = 0$ ______________________

13. $7x + y = 1$ ______________________

14. $y = 3$ ______________________

15. $x + 8 = 0$ ______________________

16. $x - 11y = 4$ ______________________

Tell whether the line (17–22) in the given graph has positive, negative, zero, or undefined slope.

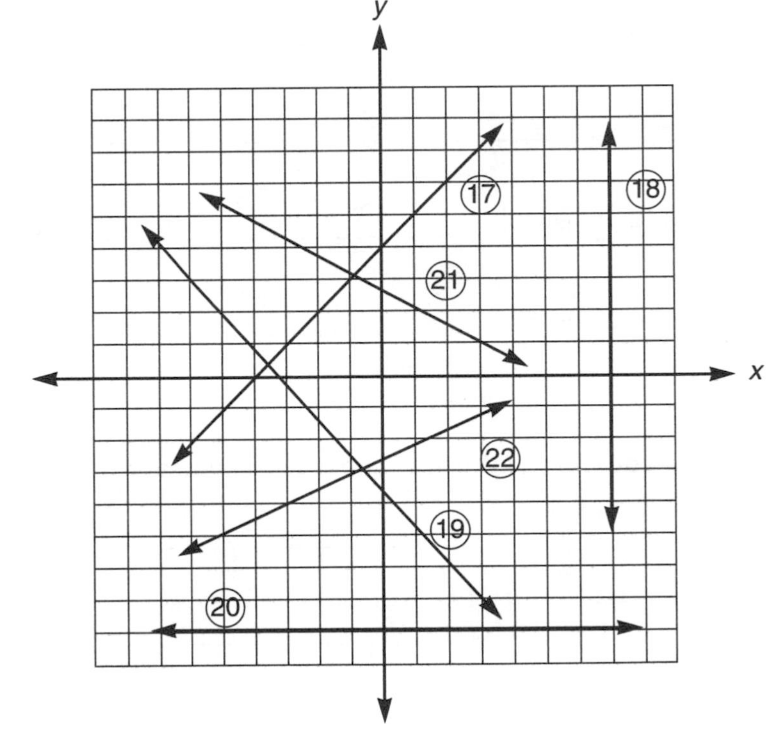

17. 

18. ______________________

19. ______________________

20. ______________________

21. ______________________

22. ______________________

Solve.

23. The vertices of a triangle are $A(2, 3)$, $B(5, ^{-}2)$, and $C(^{-}1, ^{-}4)$. Find the slope of each side of triangle ABC. ______________________

24. Without drawing the graph, tell which of these equations have graphs that pass through the point $(0, 0)$.
a. $y = 2x$ **b.** $y = 4$ **c.** $x = 6y$ **d.** $x = ^{-}3$ ______________________

25. Tell which lines have a slope equal to 5.

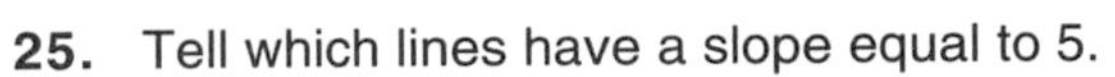

a. $2y + 5x = 1$ **b.** $3y - 15x = 9$

c. $\frac{y}{5} = x - 2$ **d.** $y = \frac{x}{5} + 5$ ______________________

*Use with Lesson 16-1, text pages 456–457.

Slope-Intercept Form*

Name ______________________

Date ______________________

Remember: Slope-Intercept Form of a linear equation is represented by:

$$y = mx + b$$

m = slope; b = point at which the line intersects the y-axis and is called the y intercept.

Find the slope and y-intercept of the line that is the graph of the equation.

1. $2x - 5y = 3$ ______________________

2. $x + y = 0$ ______________________

3. $^{-}4x + y = 7$ ______________________

4. $y + 5 = 0$ ______________________

5. $y = {^{-}2}x$ ______________________

6. $2x - 3 = 0$ ______________________

7. $\frac{1}{3}x + \frac{3}{4}y = \frac{1}{2}$ ______________________

8. $\frac{1}{5}x - \frac{1}{3} = \frac{2}{7}y$ ______________________

9. $\frac{x}{4} + 2y = 5$ ______________________

10. $\frac{x}{7} - 4y = {^{-}2}$ ______________________

Write an equation of the line whose slope and y-intercept are:

11. $m = \frac{1}{3}$; $b = 3$ ______________________

12. $m = 0$; $b = {^{-}7}$ ______________________

13. $m = {^{-}4}$; $b = 5$ ______________________

14. $m = -\frac{2}{3}$; $b = {^{-}1}$ ______________________

15. $m = {^{-}2}$; $b = 0$ ______________________

16. $m = \frac{2}{7}$; $b = \frac{1}{3}$ ______________________

State whether the lines are parallel, perpendicular or neither.

17. $2x + 5y = 6$; $y = -\frac{2}{5}x + 3$ ______________________

18. $y + 4x + 5 = 0$; $y = 4x + 5$ ______________________

19. $y = 7x$; $7y + x = 1$ ______________________

20. $y = \frac{1}{2}x + 5$; $y = -\frac{1}{2}x - 3$ ______________________

*Use with Lesson 16-2, text pages 458–459.

Finding the Equation of a Line*

Name ______________________

Date ______________________

Given: $P_1 = (2, {}^-3)$ and $P_2 = (6, {}^-1)$

Equation of line through P_1 and P_2 = ?

- $m = \frac{^-1 - {}^-3}{6 - 2} = \frac{2}{4} = \frac{1}{2}$
- $y = mx + b \longrightarrow {}^-3 = \frac{1}{2}(2) + b; b = {}^-4$
- Slope-Intercept Form: $y = \frac{1}{2}x - 4$

  or $x - 2y - 8 = 0$ (Equation of line)

Given: $m = \frac{1}{4}$ and $P = ({}^-8, 1)$

Equation of line through P = ?

- $y = mx + b \longrightarrow 1 = \frac{1}{4}({}^-8) + b; b = 3$
- Slope-Intercept Form: $y = \frac{1}{4}x + 3$

  or $x - 4y + 12 = 0$ (Equation of line)

Write the equation of the line that passes through the given points.

1. $(3, 6)$ and $(7, 9)$ ______________

2. $(0, 0)$ and $({}^-5, {}^-8)$ ______________

3. $({}^-1, {}^-5)$ and $({}^-4, {}^-9)$ ______________

4. $(4, 0)$ and $(0, {}^-7)$ ______________

5. $(4, {}^-9)$ and $({}^-5, 11)$ ______________

6. $(5, 9)$ and $(7, {}^-3)$ ______________

7. $({}^-3, {}^-3)$ and $(6, 6)$ ______________

8. $({}^-7, 3)$ and $(2, 5)$ ______________

Write the equation of the line that has the given slope and passes through the given point.

9. $m = 3$; $(4, 1)$ ______________

10. $m = {}^-6$; $({}^-2, {}^-3)$ ______________

11. $m = -\frac{1}{2}$; $(1, {}^-3)$ ______________

12. $m = -\frac{9}{5}$; $(0, {}^-5)$ ______________

13. $m = \frac{3}{5}$; $({}^-5, 0)$ ______________

14. $m = 2$; $(3, 4)$ ______________

Write the equation of the line that is:

15. parallel to the line $y = 5x - 3$ and passes through the point $({}^-1, 2)$. ______________

16. parallel to the line $3x - 5y = 2$ and passes through the point $(5, 3)$. ______________

17 perpendicular to the line $2x - 3y = 12$ and has the same y-intercept. ______________

*Use with Lesson 16-3, text pages 460–461.

Finding the Equation of a Line*
(cont'd)

Name ______________________

Date ______________________

Write the equation of the line that is:

16. perpendicular to the line $3x + 4y = 7$ and has the same y-intercept. ______________________

17. parallel to the line $x + y = {}^{-}2$ and passes through the origin. ______________________

18. perpendicular to the line $2x + y = 5$ and passes through the origin. ______________________

19. parallel to the line $y = {}^{-}2x + 7$ and has the same y-intercept as the line $y = 2x - 4$. ______________________

20. parallel to the line $y = -\frac{2}{5}x$ and has the same y-intercept as the line $2y - 3x = 5$. ______________________

Find the equation of each line (21–26) indicated in the graphs.

21. 

22. ______________________

23. 

24. 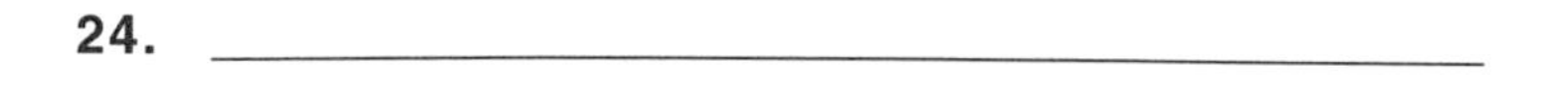

25. ______________________

26. ______________________

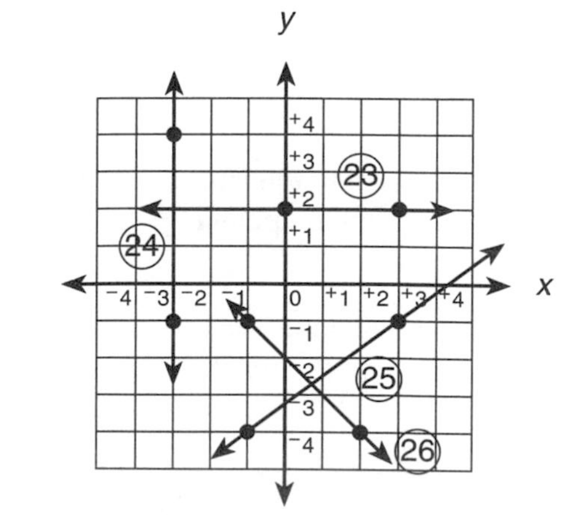

*Use with Lesson 16-3, text pages 460–461.

Equations, Slopes, and Lines*

Name ______________________

Date ______________________

Solve.

1. What is the slope of the line passing through the points $(3, 11)$ and $(7, 2)$? ______________________

2. Write an equation of the line whose slope is $^-2$ and y-intercept is 9. ______________________

3. What is the slope of the line whose equation is $\frac{1}{2}x + \frac{1}{4}y = \frac{1}{3}$? ______________________

4. What is the y-intercept of the line whose equation is $\frac{3}{5}x + \frac{1}{2} = \frac{1}{6}y$? ______________________

5. Find the equation of the line that passes through $(^-2, 5)$ and has slope of $-\frac{1}{5}$. ______________________

6. Find the equation of the line that passes through the origin and is parallel to $y = -\frac{1}{3}x + 2$. ______________________

7. What is the slope and y-intercept of the line whose equation is $5(x - y) = {^-4}$? ______________________

8. Find the slope and y-intercept of the line whose equation is $\frac{2x + 3y}{7} = 2$. ______________________

9. Find the equation of the line passing through $(^-3, 5)$ and is perpendicular to the line $2y - 5x + 2 = 0$. ______________________

10. Find the equation of the line that passes through the origin and is perpendicular to the line $3y + x - 12 = 0$. ______________________

11. Write the equation of the line passing through $(^-2, {^-2})$ and $(7, 11)$. ______________________

12. Write the equation of the line passing through $\left(\frac{1}{2}, \frac{1}{2}\right)$ and $\left(\frac{^-3}{4}, \frac{^-1}{8}\right)$. ______________________

13. A quadrilateral has vertices $A(^-2, 2)$, $B(5, 8)$, $C(7, {^-1})$, and $D(0, {^-4})$. Find the slopes of its diagonals. ______________________

*Use with Lessons 16-1, 16-2, 16-3, text pages 456–461.

Simplifying Radical Expressions*

Name ______________________

Date ______________________

Product Property of Square Roots:

$\sqrt{ab} = \sqrt{a} \cdot \sqrt{b}$ and $\sqrt{a} \cdot \sqrt{b} = \sqrt{ab}$

Quotient Property of Square Roots:

$\sqrt{\frac{a}{b}} = \frac{\sqrt{a}}{\sqrt{b}}$ and $\frac{\sqrt{a}}{\sqrt{b}} = \sqrt{\frac{a}{b}}$

Simplify.

1. $\sqrt{32}$ ______________________
2. $\sqrt{250}$ ______________________
3. $\sqrt{128x^3}$ ______________________
4. $\sqrt{75y^4}$ ______________________
5. $\sqrt{288a^3b^2}$ ______________________
6. $\sqrt{243ab^3}$ ______________________
7. $\sqrt{450x^2}$ ______________________
8. $\sqrt{625x^3y^2}$ ______________________
9. $\sqrt{512x^2y}$ ______________________
10. $\sqrt{120x}$ ______________________
11. $\sqrt{\frac{1}{2}}$ ______________________
12. $\sqrt{\frac{4}{5}}$ ______________________
13. $\sqrt{\frac{7x}{3y}}$ ______________________
14. $\sqrt{\frac{5x}{6y}}$ ______________________
15. $\sqrt{\frac{24a^7}{5c}}$ ______________________
16. $\sqrt{\frac{x}{7y}}$ ______________________
17. $\sqrt{\frac{3a}{242}}$ ______________________
18. $\sqrt{\frac{a^3bc^2}{3d}}$ ______________________
19. $\sqrt{\frac{x^2y^4}{z^3}}$ ______________________
20. $\sqrt{\frac{3y^4}{5z^3}}$ ______________________
21. $\frac{4}{\sqrt{5a}}$ ______________________
22. $\frac{6\sqrt{18}}{\sqrt{2}}$ ______________________
23. $\frac{4\sqrt{20}}{\sqrt{5}}$ ______________________
24. $\sqrt{\frac{81}{50c^2}}$ ______________________

*Use with Lesson 16-4, text pages 462–463.

Simplifying Radical Expressions* (cont'd)

Name ______________________

Date ______________________

Simplify.

25. $\dfrac{5}{\sqrt{a+b}}$ ______________

26. $\sqrt{\dfrac{15c^3d^4e^3}{3ce^4}}$ ______________

27. $\dfrac{3}{\sqrt{x-y}}$ ______________

28. $\sqrt{\dfrac{98a^3bc^2}{2ab^2}}$ ______________

29. $3x\sqrt{\dfrac{x}{3}}$ ______________

30. $5y\sqrt{\dfrac{2x}{5x^3}}$ ______________

31. $\sqrt{\dfrac{3x^3}{363y^3}}$ ______________

32. $\dfrac{\sqrt{128a^3}}{\sqrt{2a}}$ ______________

33. $xy\sqrt{\dfrac{x}{5y}}$ ______________

34. $xy\sqrt{\dfrac{x^3}{7x}}$ ______________

35. $8\sqrt{\dfrac{4a^4}{6b}}$ ______________

36. $6\sqrt{\dfrac{24x^5}{18x}}$ ______________

Tell whether each is rational or irrational.

37. $\sqrt{225}$ ______________

38. $\sqrt{12}$ ______________

39. $\sqrt{169}$ ______________

40. $\sqrt{20}$ ______________

41. $\sqrt{\dfrac{4}{16}}$ ______________

42. $\sqrt{\dfrac{5}{25}}$ ______________

43. $\frac{1}{2}\sqrt{8}$ ______________

44. $\frac{1}{3}\sqrt{9}$ ______________

***Use with Lesson 16-4, text pages 462–463.**

Addition and Subtraction of Radical Expressions*

Name ______________________

Date ______________________

Remember: Simplify the radicals if possible.
Add only like terms.

Simplify if possible.

1. $5\sqrt{3} + 6\sqrt{3}$ ______________________

2. $9\sqrt{7} + \sqrt{7}$ ______________________

3. $15\sqrt{x} - 13\sqrt{x}$ ______________________

4. $7a\sqrt{2b} - 4a\sqrt{2b}$ ______________________

5. $5\sqrt{2} + \sqrt{2} - 2\sqrt{2}$ ______________________

6. $4\sqrt{6} - \sqrt{6} - 5\sqrt{6}$ ______________________

7. $\sqrt{27} + \sqrt{12}$ ______________________

8. $\sqrt{20} + \sqrt{45}$ ______________________

9. $3\sqrt{8} - \sqrt{32}$ ______________________

10. $5\sqrt{50} - 7\sqrt{18}$ ______________________

11. $16x^2\sqrt{98} - 15x^2\sqrt{128}$ ______________________

12. $5\sqrt{108x^3} - 4x\sqrt{243x}$ ______________________

13. $\sqrt{100xy} - \sqrt{64xy} + \sqrt{25xy}$ ______________________

14. $\sqrt{300} - \sqrt{108} + 5\sqrt{48}$ ______________________

15. $\sqrt{8ab^2c} + 3b\sqrt{2ac}$ ______________________

16. $\sqrt{27a^3} + 5\sqrt{3a} - 4\sqrt{12a}$ ______________________

17. $\sqrt{\frac{11}{25}} - \sqrt{\frac{11}{49}}$ ______________________

18. $3\sqrt{\frac{5}{x^2}} + 7\sqrt{\frac{20}{x^2}}$ ______________________

19. $5\sqrt{\frac{54}{3x^4}} - 2\sqrt{\frac{216}{3x^4}}$ ______________________

20. $5\sqrt{\frac{15}{5x^2}} - \sqrt{\frac{135}{5x^2}}$ ______________________

*Use with Lesson 16-5, text pages 464–465.

Addition and Subtraction of Radical Expressions* (cont'd)

Name ______________________

Date ______________________

Simplify.

21. $\sqrt{72x^3} + 8x\sqrt{2x} - 7x\sqrt{18x}$

____

22. $4\sqrt{50} - 2\sqrt{98} + \frac{1}{2}\sqrt{72}$

____

Solve. Show your work.

23. Find the perimeter of a rectangle if the length is $\sqrt{363}$ inches and width is $\sqrt{48}$ inches. ______________________

24. What is the length of the longer leg of a right triangle if the shorter leg is 8 cm and the hypotenuse is 16 cm? ______________________

25. Find the diagonal of a rectangle 9 meters wide and 12 meters long. ______________________

26. Find the hypotenuse of an isosceles right triangle if each leg is 8 inches long. ______________________

27. If the hypotenuse of a right triangle is 25 cm and one leg is 24 cm, what would be the length of the other leg? ______________________

28. If the hypotenuse of a right triangle is 41 inches and one leg is 9 inches long, what would be the length of the other leg? ______________________

***Use with Lesson 16-5, text pages 464–465.**

Multiplication and Division of Radical Expressions*

Name ______________________

Date ______________________

To multiply radical expressions:

- Cluster similar terms.
- Simplify.
- Use the Product Rule for Radicals.

Multiply.

1. $3\sqrt{5x} \cdot 2\sqrt{5x}$ ______________

2. $4\sqrt{7} \cdot 8\sqrt{7}$ ______________

3. $5x\sqrt{2} \cdot 6x\sqrt{6}$ ______________

4. $4\sqrt{2x} \cdot 8\sqrt{3x}$ ______________

5. $^{-}7a\sqrt{a^3} \cdot 2a\sqrt{a}$ ______________

6. $\frac{1}{4}\sqrt{2b} \cdot {}^{-}20\sqrt{2b^3}$ ______________

7. $\frac{-1}{2}\sqrt{5x} \cdot {}^{-}10\sqrt{10x}$ ______________

8. $^{-}6y\sqrt{xy} \cdot {}^{-}2y\sqrt{x^2y}$ ______________

9. $\sqrt{2} \cdot \sqrt{3} \cdot \sqrt{4}$ ______________

10. $\sqrt{3} \cdot \sqrt{5} \cdot \sqrt{6}$ ______________

11. $\sqrt{27x} \cdot \sqrt{3x}$ ______________

12. $\sqrt{25a} \cdot \sqrt{ab}$ ______________

13. $\sqrt{xyz} \cdot \sqrt{4x^3yz^5}$ ______________

14. $\sqrt{a^2b^3c} \cdot \sqrt{27abc}$ ______________

15. $9\sqrt{32} \cdot 3\sqrt{2}$ ______________

16. $10\sqrt{5} \cdot \frac{1}{2}\sqrt{10}$ ______________

17. $7\sqrt{50x^3} \cdot 2\sqrt{6x}$ ______________

18. $10x\sqrt{48} \cdot 3x\sqrt{15}$ ______________

19. $9a\sqrt{a^2b} \cdot 4b\sqrt{bc}$ ______________

20. $7xy\sqrt{xy^2} \cdot 3y\sqrt{xy^3}$ ______________

21. $\sqrt{3} \cdot (\sqrt{6} - \sqrt{15})$ ______________

22. $\sqrt{ab} \cdot (\sqrt{2ab} + \sqrt{3b})$ ______________

***Use with Lesson 16-6, text pages 466–467.**

Multiplication and Division of Radical Expressions* (cont'd)

Name ______________________

Date ______________________

$$\frac{\sqrt{5}}{\sqrt{15}} = \frac{\sqrt{5}}{\sqrt{15}} \cdot \frac{\sqrt{15}}{\sqrt{15}} = \frac{\sqrt{75}}{15} = \frac{\sqrt{25 \cdot 3}}{15}$$

$$= \frac{\sqrt{25} \cdot \sqrt{3}}{15} = \frac{5\sqrt{3}}{15} = \frac{\sqrt{3}}{3}$$

$(2 + \sqrt{3}) \cdot (5 + \sqrt{3}) = \underline{\ ?\ }$

$10 + 2\sqrt{3} + 5\sqrt{3} + \sqrt{9}$

$10 + 7\sqrt{3} + 3 = 10 + 3 + 7\sqrt{3} = 13 + 7\sqrt{3}$

Compute.

23. $\frac{\sqrt{98}}{\sqrt{2}}$ ______________________

24. $\frac{\sqrt{75}}{\sqrt{3}}$ ______________________

25. $\frac{\sqrt{15}}{\sqrt{5}}$ ______________________

26. $\frac{\sqrt{42}}{\sqrt{7}}$ ______________________

27. $\frac{\sqrt{3x}}{\sqrt{2}}$ ______________________

28. $\frac{\sqrt{5y}}{\sqrt{3}}$ ______________________

29. $\frac{\sqrt{5xy}}{\sqrt{2x}}$ ______________________

30. $\frac{\sqrt{7ab}}{\sqrt{3a}}$ ______________________

31. $\frac{x\sqrt{216}}{\sqrt{3x}}$ ______________________

32. $\frac{y\sqrt{125}}{\sqrt{5y}}$ ______________________

33. $\frac{\sqrt{3} + \sqrt{2}}{\sqrt{5}}$ ______________________

34. $\frac{\sqrt{7} - \sqrt{3}}{\sqrt{2}}$ ______________________

35. $\frac{6\sqrt{12x} \cdot 3\sqrt{18y}}{2\sqrt{2}}$ ______________________

36. $\frac{7\sqrt{10a} \cdot 4\sqrt{27b}}{3\sqrt{3}}$ ______________________

37. $(7 + \sqrt{3})(2 - \sqrt{3})$ ______________________

38. $(x + \sqrt{3y})(2x + \sqrt{3y})$ ______________________

39. $(2\sqrt{2} + 5\sqrt{3})(2\sqrt{2} - 5\sqrt{3})$ ______________

40. $(4\sqrt{x} + 5\sqrt{y})(4\sqrt{x} - 5\sqrt{y})$ ______________

*Use with Lesson 16-6, text pages 466–467.

Solving Quadratic Equations by Factoring*

Name ______________________
Date ______________________

Find the solution set.

$x^2 - 9x = {}^{-}20$	Let $(x - 4) = 0$	Let $(x - 5) = 0$
$x^2 - 9x + 20 = 0$	$x = 4$	$x = 5$
$(x - 4)(x - 5) = 0$	**SS:** $\{4, 5\}$ or $x = 4$ or 5	

1. $x^2 + 9x + 14 = 0$ ______________________

2. $6x^2 - 24 = 0$ ______________________

3. $x^2 - 9x + 18 = 0$ ______________________

4. $3x^2 - 75 = 0$ ______________________

5. $x^2 + 3x - 28 = 0$ ______________________

6. $x^2 + 12x = 0$ ______________________

7. $x^2 - 6x - 55 = 0$ ______________________

8. $x^2 - 8x = 0$ ______________________

9. $x^2 + 9x = 0$ ______________________

10. $x^2 - 7x + 24 = 2x^2 - 2x$ ______________________

11. $x^2 - 15x = 0$ ______________________

12. $3x^2 - 4x - 81 = 2x^2 - 4$ ______________________

13. $x^2 + 56 = 15x$ ______________________

14. $\frac{y + 5}{5} = \frac{10}{y}$ ______________________

15. $x^2 + 48 = {}^{-}16x$ ______________________

16. $\frac{y - 7}{6} = \frac{13}{y}$ ______________________

17. $6x^2 - x - 1 = 0$ ______________________

18. $\frac{x}{75} = \frac{3}{x}$ ______________________

19. $32x^2 - 20x - 3 = 0$ ______________________

20. $\frac{x}{8} = \frac{32}{x}$ ______________________

21. $x(x + 3) = 54$ ______________________

22. $18x^2 - 15x + 2 = 0$ ______________________

23. $x(x + 7) = 60$ ______________________

24. $50x^2 - 25x + 2 = 0$ ______________________

*Use with Lesson 16-7, text pages 468–469.

Solving Quadratic Equations by Factoring* (cont'd)

Name ____________________

Date ____________________

Solve. Show your work.

25. The square of a positive number is 22 more than 9 times the number. Find the number. ____________________

26. The sum of the squares of 2 consecutive positive integers is 145. Find the integers. ____________________

27. The length of a rectangle is 5 meters more than its width. The area of the rectangle is 84 sq meters. Find the dimensions of the rectangle. ____________________

28. The perimeter of a rectangle is 30 inches and its area is 54 square inches. Find the dimensions of the rectangle. ____________________

29. The square of a number decreased by 27 is equal to 6 times the number. Find the number. ____________________

30. Find three consecutive positive integers if the product of the second and third integers is 90. ____________________

***Use with Lesson 16-7, text pages 468–469.**

Completing the Square*

Name ______________________

Date ______________________

Remember: To solve an equation by completing the square:

- The coefficient of x^2 must be equal to 1.
- Set all terms containing x on one side of the equation.
- Take $\frac{1}{2}$ of the coefficient of x, square it, and add it to both sides of the equation.
- Factor. Then take the square root of both sides to solve for x.

Solve by completing the square.

1. $x^2 + x - 1 = 0$ ______________________

2. $x^2 - 3 = 5x$ ______________________

3. $2x^2 + 8x - 12 = 0$ ______________________

4. $x^2 + 2x = 4$ ______________________

5. $3x^2 + 9x - 6 = 0$ ______________________

6. $x^2 - 4x + 1 = 0$ ______________________

7. $9x^2 + 3 = 36x$ ______________________

8. $7x^2 - 49 = 21x$ ______________________

9. $4x^2 = 8x + 1$ ______________________

10. $11x^2 - 121 = 22x$ ______________________

Solve.

11. The difference between a positive number and its reciprocal is $\frac{5}{6}$. Find the number. ______________________

12. The square of a number decreased by 7 is equal to 1 more than 7 times the number. Find the number. ______________________

13. The length of the base of a parallelogram is twice its height. The area of the parallelogram is 36 square centimeters. Find the length of its base and its height. ______________________

14. The height of a triangle measures 3 cm less than its base. The area of the triangle is 20 cm^2. Find the lengths of its base and height. ______________________

*Use with Lesson 16-8, text pages 470–471.

Quadratic Formula*

Name ______________________

Date ______________________

Remember:

If $ax^2 + bx + c = 0$, and a, b, and c are real numbers and $a \neq 0$, then:

$$x = \frac{-b \pm \sqrt{b^2 - 4ac}}{2a}$$

Solve. Use the Quadratic Formula.

1. $x^2 - 2x - 10 = 0$ ______________

2. $x^2 + 14 = 8x$ ______________

3. $x^2 + 6x + 2 = 0$ ______________

4. $x^2 + 19 = 10x$ ______________

5. $4x^2 - 4x - 1 = 0$ ______________

6. $4x^2 + x = 3x^2 + x + 21$ ______________

7. $9x^2 + 12x + 1 = 0$ ______________

8. $5x^2 + x - 20 = 4x^2 + x + 5$ ______________

9. $16x^2 - 24x - 4 = 0$ ______________

10. $2x^2 + 5x = 1$ ______________

Solve. (Hint: Some problems may have more than one answer.)

11. One positive number is 3 more than twice another. Their product is 119. Find the numbers. ______________

12. One positive number is 9 less than another. The sum of the squares of these numbers is 153. Find the numbers. ______________

13. The sum of a number and the square of its additive inverse is 72. Find the number. ______________

14. The sides of a rectangle are represented by x and $4x + 3$. The area of the rectangle is 76 cm^2. Find the lengths of the sides. ______________

***Use with Lesson 16-9, text pages 472–473.**

Solving Quadratic Equations*

Name ______________________

Date ______________________

Solve. (Use any method.)

1. $x^2 - 4x - 21 = 0$ ______________

2. $x^2 + 5x - 66 = 0$ ______________

3. $10x^2 - 11x + 3 = 0$ ______________

4. $6x^2 - 19x + 10 = 0$ ______________

5. $x^2 - 6x - 5 = 0$ ______________

6. $x^2 - 10x + 14 = 0$ ______________

7. $3x^2 - 2 = x$ ______________

8. $4x^2 + 9x = 9$ ______________

9. $2x^2 - 12x = 0$ ______________

10. $3x^2 = 72$ ______________

11. $4x^2 - 12x = 112$ ______________

12. $5x^2 - 175 = 10x$ ______________

13. $8x^2 + 12 = {}^-20x$ ______________

14. $12x^2 + 30 = 39x$ ______________

15. $20x^2 - 10x = 5$ ______________

16. $18x^2 + 2 = 24x$ ______________

17. $\frac{2x}{3} = \frac{150}{x}$ ______________

18. $\frac{x}{10} = \frac{32}{5x}$ ______________

*Use with Lesson 16-9, text pages 472–473.

Solving Quadratic Equations* (cont'd)

Name ______________________

Date ______________________

Remember: Quadratic equations are not always expressed in terms of x.

19. $\frac{c + 3}{5} = \frac{2}{c}$ ______________________

20. $\frac{2r + 15}{2} = \frac{4}{r}$ ______________________

21. $4n^2 + 1 = {}^{-}4n$ ______________________

22. $e^2 + 1 = 3e$ ______________________

23. $30y^2 = 2 + 7y$ ______________________

24. $21y^2 + 5y - 6 = 0$ ______________________

25. $y^2 = 4y + 14$ ______________________

26. $2y^2 - 6y + 3 = 0$ ______________________

Solve.

27. Twice the square of a number decreased by three times the number is 9. Find the number. ______________________

28. The larger of two positive numbers is 4 more than the smaller. The product of the numbers is 5. Find the numbers. ______________________

29. 16 times a certain number is 5 more than 3 times the square of the number. Find the number. ______________________

30. The sum of two numbers is 4 and their product is $^{-}45$. Find the numbers. ______________________

31. One leg of a right triangle is 7 inches longer than the other leg. The hypotenuse is 17 in. Find the length of each leg of the triangle. ______________________

32. A rectangle has an area of 25 square units. The sides of the rectangle are represented by x and $4x - 15$. Find the lengths of these sides. ______________________

*Use with Lesson 16-9, text pages 472–473.

Advanced Placement Review*

Name ____________________

Date ____________________

Circle the correct answer.

1. The value of $16a^2b^2 - 9b$ when $a = -1$ and $b = 2$ is:

a. -82 **b.** -78 **c.** 46 **d.** 42

2. Subtract the sum of $5a^2 + 14$ and $-9a + 7$ from $7a^2 - 4a - 3$. The result is:

a. $2a^2 + 5a - 24$ **b.** $-2a^2 - 5a + 24$ **c.** $2a^2 - 5a - 24$ **d.** $-2a^2 + 5a - 24$

3. $x^2y(3x^2 - 5y^2)$ is equal to:

a. $3x^4 - 5xy^3$ **b.** $3x^4y - 5x^2y^3$ **c.** $3x^4y - 5xy^3$ **d.** $3x^4 - 5xy^3$

4. Which of the following is prime?

a. $16x^2 + 64y^2$ **b.** $16x^2 - 64y^2$ **c.** $9x^2 - 24xy + 16y^2$ **d.** $9x^2 - 12xy + 16y^2$

5. The least common denominator of $\frac{7x + 2}{3x - 6}$ and $\frac{5x - 1}{2x - 4}$ is:

a. $(3x - 6)(2x - 4)$ **b.** $6(x - 2)$ **c.** $3(x - 2)$ **d.** $2(x - 2)$

6. If $m = \frac{-1}{3}$ and $b = 2$, the equation of the line is:

a. $x + 3y - 2 = 0$ **b.** $x + 3y - 6 = 0$ **c.** $x - 3y + 2 = 0$ **d.** $x - 3y - 6 = 0$

7. $\frac{3}{a^2 + 6a} \div \frac{9}{a + 6}$ is equal to:

a. $\frac{1}{3a}$ **b.** $\frac{1}{9a}$ **c.** $\frac{a + 6}{3}$ **d.** $\frac{a + 6}{9}$

8. Which of the following is an irrational number?

a. $\sqrt{400}$ **b.** $\sqrt{625}$ **c.** $-\sqrt{54}$ **d.** $-\sqrt{36}$

9. $4\sqrt{48} - 8\sqrt{12}$ is equal to:

a. $8\sqrt{3}$ **b.** 0 **c.** $-4\sqrt{3}$ **d.** $-2\sqrt{3}$

10. $\frac{16\sqrt{21}}{2\sqrt{7}}$ is equal to:

a. $24\sqrt{3}$ **b.** $8\sqrt{7}$ **c.** $8\sqrt{3}$ **d.** 24

11. The solution set of the equation $x^2 - 15x + 56 = 0$ is:

a. $\{-7, -8\}$ **b.** $\{7, 8\}$ **c.** $\{7, -8\}$ **d.** $\{-7, 8\}$

12. The solution set of the equation $x^2 - 2x - 2 = 0$ is:

a. $\left\{1 \pm \sqrt{3}\right\}$ **b.** $\left\{^-1 \pm \sqrt{3}\right\}$ **c.** $\left\{\frac{2 \pm \sqrt{3}}{2}\right\}$ **d.** $\left\{\frac{^-2 \pm \sqrt{3}}{2}\right\}$

13. The diagonal of a square whose side is 5 cm is:

a. $5\sqrt{2}$ cm **b.** 10 cm **c.** $25\sqrt{2}$ cm **d.** $2\sqrt{10}$ cm

Solve.

14. The sum of the reciprocals of two numbers is $\frac{5}{18}$.
If one number is three less than the other, find the numbers. ____________________

15. A 25 ft-ladder leaning against the wall reaches a point 15 ft above the ground. How far is the foot of the ladder from the foot of the wall? ____________________

*Use to review Chapters 15 and 16.

Nonroutine Problems*

Name ______________________________

Date ______________________________

Solve.

1. Speedy Spike and Motor Mike are racers who are racing in the same direction. Speedy Spike left 2 hours before Motor Mike and is going 70 mph. Motor Mike is going 80 mph. How many miles apart will they be 30 minutes before the faster one catches the slower one?

2. A swimmer does 40 laps (down and back) in a square pool 2025 square feet in area. How many yards does the swimmer swim?

3. Directions on a frozen-juice can 6.2 cm in diameter and 10.5 cm deep recommend adding 2 cans of water to the concentrate. To the nearest milliliter, what will the total volume of juice be when this is done?

4. A baby looks up a flight of stairs. His toy is on the fifth step. Each step is 12 inches deep and 9 inches high. If the baby could travel in a straight line, what is the distance he would travel from the floor to the bottom of the fifth step?

5. The perimeter of a rectangle is 72 feet. If the length exceeds 4 times the width by 1 foot, what are the dimensions of the rectangle?

*Use for Chapters 11 through 16.

Advanced Placement-Type Test*

Name ______________________

Date ______________________

Circle the correct answer.

1. $(3x^2)^3$ is equivalent to:
a. $3x^5$ **b.** $27x^5$ **c.** $27x^6$ **d.** $3x^6$

2. The sum of $x^2 + 5x - 24$ and $-4x^2 - 12x + 5$ is:
a. $-3x^2 - 7x - 19$ **b.** $3x^2 - 7x - 16$ **c.** $-3x^2 + 17x - 18$ **d.** $3x^2 + 17x - 29$

3. Subtract $2a^2 - 3a + 7$ from $a^2 + 6a - 12$. The result is:
a. $3a^2 - 3a - 5$ **b.** $a^2 + 3a - 5$ **c.** $-a^2 + 9a + 19$ **d.** $-a^2 + 9a - 19$

4. $^{-}x^3y^3\,(6x^4y - 3y^4)$ is equal to:
a. $-6x^7y^4 + 3x^3y^7$ **b.** $-6x^{12}y^3 + 3x^3y^{12}$ **c.** $6x^7y^4 + 3x^3y^7$ **d.** $6x^{12}y^3 + 3x^3y^{12}$

5. $(5a + 2b)\,(3a + 4b)$ is equal to:
a. $15a^2 + 26ab + 6b^2$ **b.** $8a^2 + 26ab + 8b^2$ **c.** $15a^2 + 15ab + 8b^2$ **d.** $15a^2 + 26ab + 8b^2$

6. Evaluate $6a^4b - 5a^2b^3$ when **a** $= -1$ and **b** $= 2$. The result is:
a. 28 **b.** –28 **c.** –42 **d.** 30

7. $(9x^3 - 6x^2 + 3x) \div 3x$ is equal to:
a. $3x^2 + 2x + 1$ **b.** $3x^2 - 2x + 1$ **c.** $3x^3 + 2x + 1$ **d.** $3x^3 - 2x + 1$

8. $xy - [x^2 - x\,(y + x)]$ is equal to:
a. $2x^2 - 2xy$ **b.** $2x^2 - xy$ **c.** $2xy$ **d.** $2x^2$

9. In factored form, $8x^3y^2z^3 + 12x^2y^2z^2$ is:
a. $2x^2y^2z\,(4z^2 + 3x)$ **b.** $2x^2y^2z^2\,(4xz + 3)$ **c.** $4xyz^2\,(2x^2y + 3z)$ **d.** $4x^2y^2z^2\,(2xz + 3)$

10. As a product of 2 binomials, $2x^2 - 9xy - 5y^2$ is equal to:
a. $(2x + y)\,(x - 5y)$ **b.** $(2x - y)\,(x + 5y)$ **c.** $(2x + 5y)\,(x - y)$ **d.** $(2x - 5y)\,(x + y)$

11. If the measure of a side of a square is $3x - 2y$, the trinomial that represents the area of the square is:
a. $9x^2 + 4xy + 4y^2$ **b.** $9x^2 - 6xy + 4y^2$ **c.** $9x^2 + 6xy + 4y^2$ **d.** $9x^2 - 12xy + 4y^2$

12. Which of the following is prime?
a. $x^2 - y^2$ **b.** $8x^2 + 16y^2$ **c.** $x^2 - xy + y^2$ **d.** $x^2 + 2xy + y^2$

13. When factored completely, $x^4 + x^2 - 2$ is equal to:
a. $(x^2 + 2)\,(x + 1)\,(x - 1)$ **b.** $(x^2 + 2)\,(x^2 - 1)$
c. $(x^2 + 1)\,(x^2 - 2)$ **d.** $(x^2 + 1)\,(x + 2)\,(x - 2)$

*Use after Chapters 15, 16.

Advanced Placement-Type Test*

Name ______________________

Date ______________________

14. For what value of x is $\frac{x + 5}{x + 2}$ undefined?

a. $^-5$ **b.** $^-2$ **c.** 0 **d.** no value

15. The sides of a triangle are represented by $\frac{x}{2}$, $\frac{4x}{5}$, and $\frac{5x}{6}$. Find the perimeter of the triangle.

a. $\frac{63x}{64}$ **b.** $\frac{31x}{15}$ **c.** $\frac{x}{2}$ **d.** $\frac{32x}{15}$

16. $\frac{a^2 - 5a}{a^2} \cdot \frac{a}{2a - 10}$ is equal to:

a. $\frac{1}{2}$ **b.** $\frac{a}{2}$ **c.** $\frac{a - 5}{2}$ **d.** $\frac{a}{a - 5}$

17. The least common denominator of $\frac{3x - 2}{2x + 2}$ and $\frac{4x - 1}{3x + 3}$ is:

a. $(2x + 2)(3x + 3)$ **b.** $3(x + 1)$ **c.** $6(x + 1)$ **d.** $2(x + 1)$

18. What value of k will complete the square of the equation $x^2 - 18x + k = 0$?

a. 9 **b.** 81 **c.** $^-81$ **d.** 36

19. The slope of the line passing through $\left(\frac{^-1}{2}, \frac{^-1}{3}\right)$, and $\left(\frac{1}{4}, \frac{^-5}{6}\right)$ is:

a. $\frac{^-7}{9}$ **b.** $\frac{^-1}{8}$ **c.** $\frac{^-2}{3}$ **d.** $\frac{^-3}{8}$

20. The y-intercept of the line $5x + 3y - 9 = 0$ is:

a. 3 **b.** $\frac{5}{3}$ **c.** $^-3$ **d.** $\frac{^-3}{8}$

21. The equation of the line passing through the origin and has slope $\frac{2}{5}$ is:

a. $2x + 5y = 0$ **b.** $2x - 5y = 0$ **c.** $5x - 2y = 0$ **d.** $5x + 2y = 0$

22. Which line is parallel to $7x - 9y + 1 = 0$?

a. $x + 9y - 7 = 0$ **b.** $7x + y - 9 = 0$ **c.** $7x - 9y = 0$ **d.** $7x + 9y + 2 = 0$

23. $\sqrt{125} + 8\sqrt{5}$ can be written in the form of $x\sqrt{5}$. Find the value of x.

a. 13 **b.** 33 **c.** 10 **d.** 23

24. The product of $\sqrt{7xy}$ and $\sqrt{28xy}$ is:

a. $7xy\sqrt{2}$ **b.** $14xy$ **c.** $2xy\sqrt{7}$ **d.** $8xy$

***Use after Chapters 15, 16.**

Advanced Placement-Type Test*

Name ______________________

Date ______________________

25. The solution set of $3x^2 = 7 - 4x$ is:

a. $\left\{1, \frac{7}{3}\right\}$ **b.** $\left\{^-1, \frac{^-7}{3}\right\}$ **c.** $\left\{^-1, \frac{7}{3}\right\}$ **d.** $\left\{1, \frac{^-7}{3}\right\}$

26. $\sqrt{\frac{x + y}{x - y}}$ is equivalent to:

a. $\frac{\sqrt{x^2 - y^2}}{x - y}$ **b.** $\frac{\sqrt{x + y}}{x - y}$ **c.** $\frac{(x + y)\sqrt{x - y}}{x - y}$ **d.** $\frac{\sqrt{x - y}}{x + y}$

27. The solution set of $x^2 = 2x + 6$ is:

a. $\left\{\frac{1 \pm \sqrt{7}}{2}\right\}$ **b.** $\left\{\frac{^-1 \pm \sqrt{7}}{2}\right\}$ **c.** $\left\{1 \pm \sqrt{7}\right\}$ **d.** $\left\{^-1 \pm \sqrt{7}\right\}$

28. Between what two positive numbers does $2\sqrt{13}$ lie?

a. 6 and 7 **b.** 7 and 8 **c.** 8 and 9 **d.** 9 and 10

29. The sum of the square of a positive integer and 5 is 230. Find the positive integer.

a. 5 **b.** 12 **c.** 23 **d.** 15

30. What is the square root of $81x^2 - 144xy + 64y^2$?

a. $9x - 8y$ **b.** $9x + 8y$ **c.** $9x^2 - 8y^2$ **d.** $9x^2 + 8y^2$

31. Which points form a line that has a slope that is undefined?

a. $(4, 7)$ and $(^-3, 7)$ **b.** $(3, ^-2)$ and $(3, ^-5)$
c. $(^-1, ^-1)$ and $(^-2, ^-2)$ **d.** $(2, 2)$ and $(5, 5)$

32. What is the height of an equilateral triangle that has 12 cm as the length of its side?

a. $6\sqrt{2}$ cm **b.** $6\sqrt{3}$ cm **c.** 6 **d.** $12\sqrt{3}$ cm

33. The lengths of the legs of a right triangle are both 5 cm. Find the length of the hypotenuse.

a. $5\sqrt{2}$ cm **b.** $10\sqrt{2}$ cm **c.** $\frac{5\sqrt{2}\text{ cm}}{2}$ **d.** 10 cm

34. Which of the following is a Pythagorean triple?

a. 4, 5, 6 **b.** $4, 4\sqrt{2}, 8$ **c.** $3\sqrt{2}, 4\sqrt{2}, 5\sqrt{2}$ **d.** $6, 6, 6\sqrt{3}$

35. If $5a + b$ represents the width of a rectangle and $7a + 2b$ represents the length, then which of the following represents the perimeter?

a. $13a + 3b$ **b.** $12a + 3b$ **c.** $26a + 6b$ **d.** $24a + 6b$

*Use after Chapters 15, 16.

Advanced Placement-Type Test*

Name ______________________

Date ______________________

36. If $9x - y$ represents the radius of a circle, which of the following represents the circumference?

a. $(9x^2 - y^2)\pi$ **b.** $(18x - y)\pi$ **c.** $(18x - 2y)\pi$ **d.** $(9x - y)\pi$

37. If x represents the second of 3 consecutive integers, which expression represents the largest integer?

a. $x + 1$ **b.** $x - 1$ **c.** $x + 2$ **d.** $x - 2$

38. If one factor of the product $4x^2 - 5xy - 6y^2$ is $x - 2y$, then the other factor is:

a. $2x + 6y$ **b.** $2x - 6y$ **c.** $4x + 3y$ **d.** $4x - 3y$

39. $81x^2 + 18x + 1$ represents the area of a square. The binomial that represents the length of a side is:

a. $9x - 1$ **b.** $9x + 1$ **c.** $81x + 1$ **d.** $81x - 1$

40. Which line is perpendicular to $7x - 5y = 0$?

a. $5x - 7y = 0$ **b.** $^{-}7x - 5y = 0$ **c.** $5x + 7y = 0$ **d.** $7x + 5y = 0$

Compute.

41. $\frac{6}{x^2y} + \frac{3}{xy^2} + \frac{4}{xy}$ ______________

42. $\frac{6x + 18}{x^2 - 9} \div \frac{x + 3}{x - 3} \cdot \frac{x + 3}{6}$ ______________

43. $\frac{\sqrt{6x} \cdot \sqrt{2x}}{\sqrt{x^3 + 3x}}$ ______________

44. $(5\sqrt{7} + 4\sqrt{3})(6\sqrt{7} - 2\sqrt{3})$ ______________

45. $\frac{a + 2}{a^2 - 2a - 15} - \frac{4a - 3}{a^2 - 7a + 10} + \frac{2a - 1}{a^2 + a - 6}$ ______________

Solve.

46. The sum of the reciprocals of two positive numbers is $\frac{1}{2}$. One number is 3 more than the other. Find the numbers. ______________

47. Ann has $1.70 in dimes and nickels. There are 8 more dimes than nickels. Find the number of each kind of coin that she has. ______________

48. The numerator of a fraction is 12 less than the denominator of the fraction. The value of the fraction is $\frac{7}{11}$. Find the fraction. ______________

49. The hypotenuse of a right triangle is 4 cm longer than one leg and 18 cm longer than the other leg. Find the length of each side of the triangle. ______________

50. The smaller of two pipes takes 5 hours longer than the larger to fill a tank. If both pipes are used, the job can be done in $3\frac{1}{3}$ hours. How long will it take each pipe to fill the tank alone? ______________

Score: 2 points each.

Tables for Measures

Metric Units

Length

1 millimeter (mm) = 0.001 meter (m)
1 centimeter (cm) = 0.01 meter
1 decimeter (dm) = 0.1 meter

1 dekameter (dam) = 10 meters
1 hectometer (hm) = 100 meters
1 kilometer (km) = 1000 meters

Mass

1 milligram (mg) = 0.001 gram (g)
1 kilogram (kg) = 1000 grams
1 metric ton (t) = 1000 kilograms

Capacity

1 milliliter (mL) = 0.001 liter (L)
1 kiloliter (kL) = 1000 liters

Temperature

0° Celsius (C) .... Water freezes.
100° Celsius (C) .... Water boils.

Customary Units

Length

1 foot (ft) = 12 inches (in.)
1 yard (yd) = 36 inches
1 yard (yd) = 3 feet
1 mile (mi) = 5280 feet
1 mile (mi) = 1760 yards

Weight

1 pound (lb) = 16 ounces (oz)
1 ton = 2000 pounds

Capacity

3 teaspoons (tsp) = 1 tablespoon (tbsp)
1 cup (c) = 8 fluid ounces (fl oz)
1 pint (pt) = 2 cups
1 quart (qt) = 2 pints
1 quart (qt) = 4 cups
1 gallon (gal) = 4 quarts

Temperature

32° Fahrenheit (F) ... Water freezes.
212° Fahrenheit (F) ... Water boils.

Mathematical Symbols

$=$	is equal to
$\neq$	is not equal to
$>$	is greater than
$<$	is less than
$\geq$	is greater than or equal to
$\leq$	is less than or equal to
ϕ, $\{\}$	the empty set
...	continues without end
$\subset$	is a subset of
$\cup$	union
$\cap$	intersection
$\wedge$	conjunction
$\vee$	disjunction
$n!$	factorial $n \cdot (n-1) \cdot (n-2) \cdot \ldots \cdot 1$
10^2	ten squared
$0.\overline{3}$	0.333 ... (repeating decimals)
%	percent
$^\circ$	degree
$\cdot$	times
$\overleftrightarrow{AB}$	line AB
$\overline{AB}$	segment AB
$\overrightarrow{AB}$	ray AB
$\angle ABC$	angle ABC
ABC	plane ABC
$\sim$	is similar to
$\cong$	is congruent to
$\parallel$	is parallel to
$\perp$	is perpendicular to
2 : 3	two to three (ratio)
π	pi
$\approx$	is approximately equal to
(3, 4)	ordered pair
$P(E)$	probability of an event
$\widehat{AB}$	arc AB

Geometric Formulas

Perimeter

Rectangle: $P = 2(\ell + w)$

Square: $P = 4s$

Area

Rectangle: $A = \ell w$

Square: $A = s^2$

Parallelogram: $A = bh$

Triangle: $A = \frac{1}{2} bh$

Trapezoid: $A = \frac{1}{2}(b_1 + b_2)h$

Circle: $A = \pi r^2$

Pythagorean Theorem: $c^2 = a^2 + b^2$

Circumference of Circle

$C = \pi d$ or $2\pi r$

Surface Area

Rectangular Prism:

$S = 2(\ell w + \ell h + wh)$

Cube: $S = 6s^2$

Volume

Prism: $V = Bh$

Cube: $V = e^3$

Pyramid: $V = \frac{1}{3} Bh$

Other Formulas

Distance = Rate × Time: $d = r \times t$

Discount = List Price × Rate of Discount: $D = LP \times R$ of D

Sales Tax = Marked Price × Rate of Sales Tax: $T = MP \times R$ of T

Commission = Total Sales × Rate of Commission: $C = TS \times R$ of C

Interest = Principal × Rate × Time: $I = P \times R \times T$

Slope: $(y_2 - y_1) \div (x_2 - x_1)$

Table of Trigonometric Ratios

Angle	Sin	Cos	Tan	Angle	Sin	Cos	Tan
0°	0.000	1.000	0.000	**45°**	0.707	0.707	1.000
1°	0.017	1.000	0.017	**46°**	0.719	0.695	1.036
2°	0.035	0.999	0.035	**47°**	0.731	0.682	1.072
3°	0.052	0.999	0.052	**48°**	0.743	0.669	1.111
4°	0.070	0.998	0.070	**49°**	0.755	0.656	1.150
5°	0.087	0.996	0.087	**50°**	0.766	0.643	1.192
6°	0.105	0.995	0.105	**51°**	0.777	0.629	1.235
7°	0.122	0.993	0.123	**52°**	0.788	0.616	1.280
8°	0.139	0.990	0.141	**53°**	0.799	0.602	1.327
9°	0.156	0.988	0.158	**54°**	0.809	0.588	1.376
10°	0.174	0.985	0.176	**55°**	0.819	0.574	1.428
11°	0.191	0.982	0.194	**56°**	0.829	0.559	1.483
12°	0.208	0.978	0.213	**57°**	0.839	0.545	1.540
13°	0.225	0.974	0.231	**58°**	0.848	0.530	1.600
14°	0.242	0.970	0.249	**59°**	0.857	0.515	1.664
15°	0.259	0.966	0.268	**60°**	0.866	0.500	1.732
16°	0.276	0.961	0.287	**61°**	0.875	0.485	1.804
17°	0.292	0.956	0.306	**62°**	0.883	0.469	1.881
18°	0.309	0.951	0.325	**63°**	0.891	0.454	1.963
19°	0.326	0.946	0.344	**64°**	0.899	0.438	2.050
20°	0.342	0.940	0.364	**65°**	0.906	0.423	2.145
21°	0.358	0.934	0.384	**66°**	0.914	0.407	2.246
22°	0.375	0.927	0.404	**67°**	0.921	0.391	2.356
23°	0.391	0.921	0.424	**68°**	0.927	0.375	2.475
24°	0.407	0.914	0.445	**69°**	0.934	0.358	2.605
25°	0.423	0.906	0.466	**70°**	0.940	0.342	2.747
26°	0.438	0.899	0.488	**71°**	0.946	0.326	2.904
27°	0.454	0.891	0.510	**72°**	0.951	0.309	3.078
28°	0.469	0.883	0.532	**73°**	0.956	0.292	3.271
29°	0.485	0.875	0.554	**74°**	0.961	0.276	3.487
30°	0.500	0.866	0.577	**75°**	0.966	0.259	3.732
31°	0.515	0.857	0.601	**76°**	0.970	0.242	4.011
32°	0.530	0.848	0.625	**77°**	0.974	0.225	4.332
33°	0.545	0.839	0.649	**78°**	0.978	0.208	4.705
34°	0.559	0.829	0.675	**79°**	0.982	0.191	5.145
35°	0.574	0.819	0.700	**80°**	0.985	0.174	5.671
36°	0.588	0.809	0.727	**81°**	0.988	0.156	6.314
37°	0.602	0.799	0.754	**82°**	0.990	0.139	7.115
38°	0.616	0.788	0.781	**83°**	0.993	0.122	8.144
39°	0.629	0.777	0.810	**84°**	0.995	0.105	9.514
40°	0.643	0.766	0.839	**85°**	0.996	0.087	11.430
41°	0.656	0.755	0.869	**86°**	0.998	0.070	14.301
42°	0.669	0.743	0.900	**87°**	0.999	0.052	19.081
43°	0.682	0.731	0.933	**88°**	0.999	0.035	28.636
44°	0.695	0.719	0.966	**89°**	1.000	0.017	57.290
45°	0.707	0.707	1.000	**90°**	1.000	0.000	———

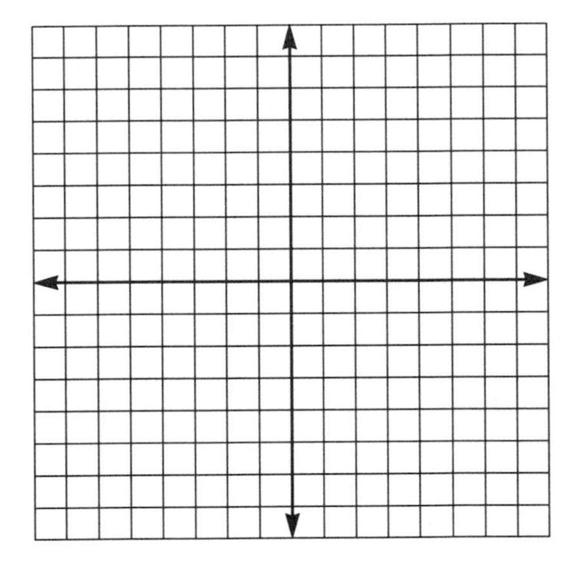

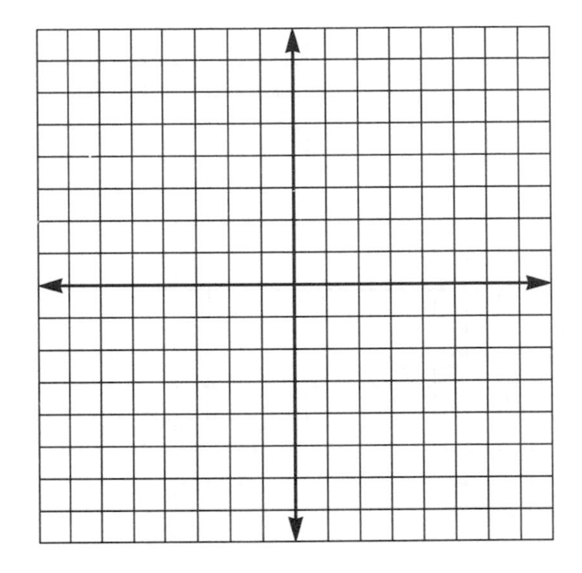

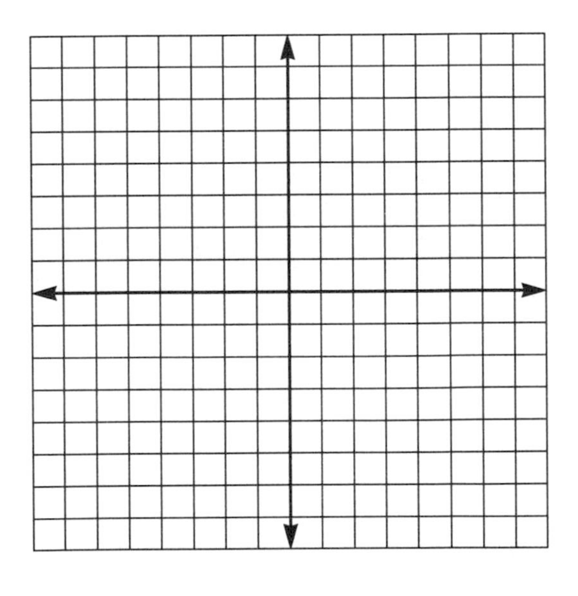

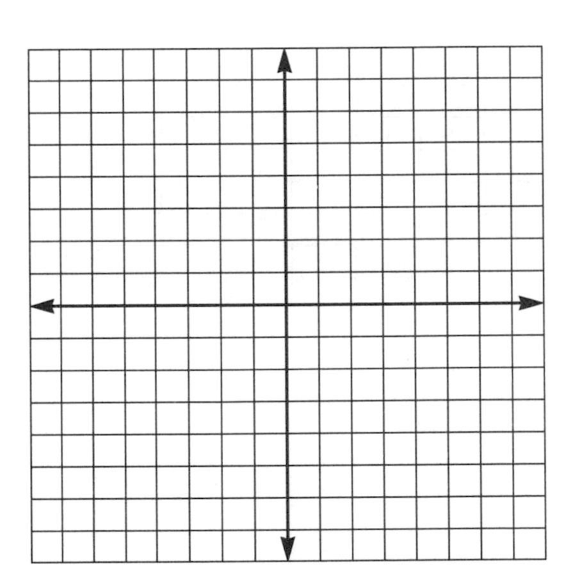

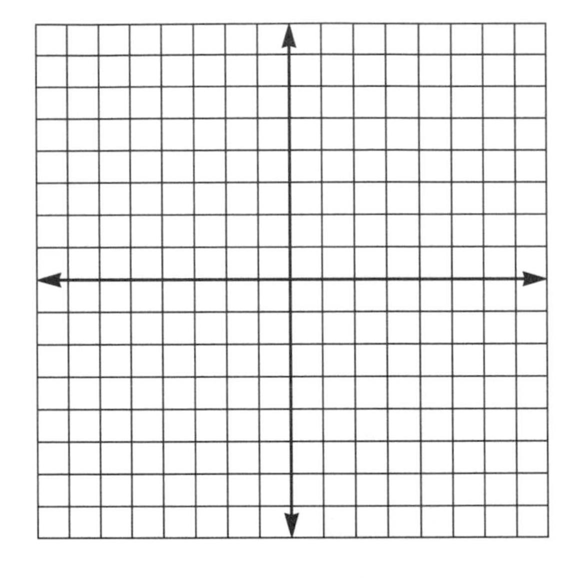

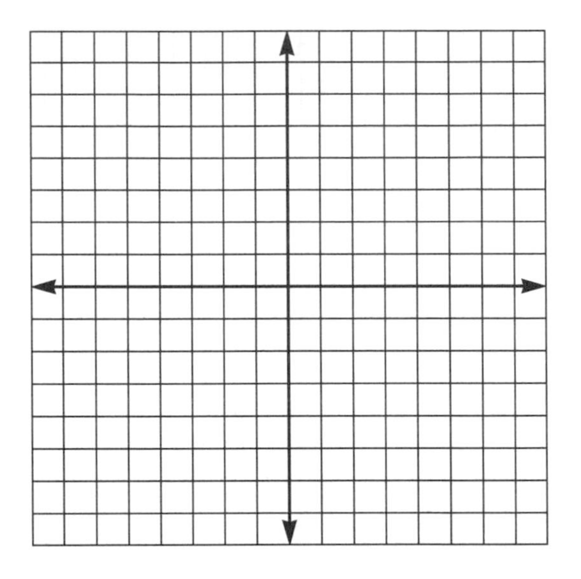

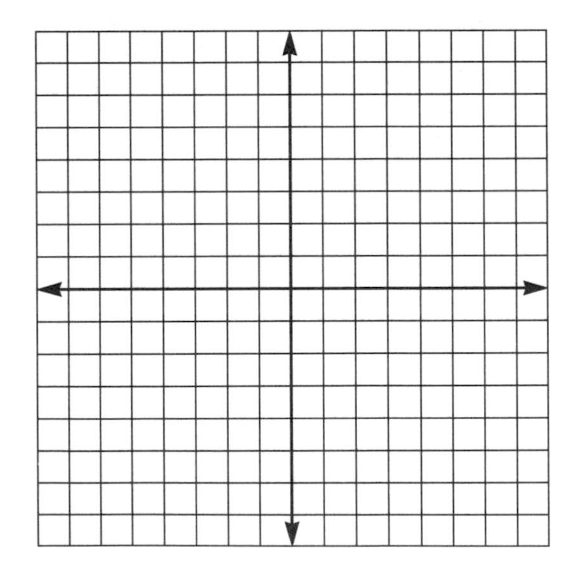

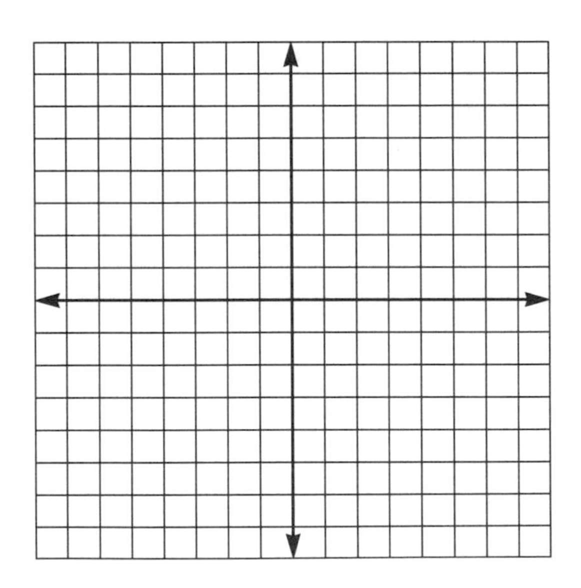

Answers to Selected Odd-Numbered Exercises

Page 3: **1.** 11,640,018 **3.** 10,000,500,002,000 **5.** ten millions; zero ten million **7.** ten thousands; 7 ten thousands **9.** billions; 2 billions **11.** hundred billions; 9 hundred billions **13.** one billion times greater **15.** 10 billion times greater **17.** 114,000,000; no; no

Page 4: **1.** zero hundredths **3.** 7 tenths **5.** 7 millionths **7.** 80,402 **9.** 0.7008 **11.** 6,000,000 **13.** 7,000,000,000 **15.** 4,200,000,000.2 **17.** 3,800,000,000,000 **19.** 3.600508 **21.** 0.022908006 **23.** 98,760 **25.** 1.083052

Page 5: **1.** 1.6987 **3.** 2.4614 **5.** 5.89 **7.** 38.421 **9.** 1.8015 **11.** 2.7038 **13.** 14.9693 **15.** 2.2 **17.** 1.9 **19.** 402

Page 6: **1.** 3,618,318 **3.** 1,330,596 **5.** 2,976,137 **7.** 1,943,980 **9.** 1,877,720 **11.** 4,206,736 **13.** 5,400,000 **15.** 32,000 **17.** 90,000 **19.** 15,000,000 **21.** 30,000,000

Page 7: **1.** 201.24 **3.** 32.0492 **5.** 731.3141 **7.** 1.72414 **9.** 12.714 **11.** 15.0627 **13.** 74.9115 **15.** 27.9097 **17.** 360 **19.** 0.36 **21.** 160 **23.** $8.14

Page 8: **1.** 780 **3.** 9020 R30 **5.** 407 R10 **7.** 600 R9 **9.** 951 **11.** 680 **13.** 36 R12 **15.** 68 **17.** 463 R67 **19.** 50 R23 **21.** 50 **23.** 800 **25.** 20

Page 9: **1.** 0.16 **3.** 3.4 **5.** 20.3 **7.** 16.87 **9.** 20.03 **11.** 1.86 **13.** 0.05 **15.** 50 **17.** 2 **19.** 0.41L

Page 10: **1.** = **3.** > **5.** = **7.** 107,500 **9.** 30.2 **11.** 560,000 **13.** 200 **15.** 9 **17.** R **19.** R

Page 11: **1.** e **3.** h **5.** g **7.** j **9.** i **11.** 490 **13.** 8.7 **15.** 300 **17.** = **19.** = **21.** <

Page 12: **1.** $\frac{16}{3}$ **3.** $\frac{73}{9}$ **5.** $\frac{11}{3}$ **7.** $\frac{17}{8}$ **9.** $\frac{71}{5}$ **11.** $\frac{48}{5}$ **13.** $\frac{67}{5}$ **15.** $\frac{50}{3}$ **17.** $\frac{125}{6}$ **19.** $\frac{64}{5}$ **21.** $\frac{84}{5}$ **23.** $\frac{99}{8}$ **25.** $9\frac{2}{3}$ **27.** $5\frac{2}{3}$ **29.** $5\frac{1}{7}$ **31.** $11\frac{5}{7}$ **33.** $5\frac{1}{6}$ **35.** $9\frac{4}{5}$ **37.** $7\frac{3}{7}$ **39.** $9\frac{1}{3}$ **41.** $22\frac{3}{4}$ **43.** $12\frac{1}{3}$ **45.** $40\frac{1}{2}$ **47.** $14\frac{1}{3}$ **49.** $\frac{15}{4}$ **51.** $\frac{8}{3}$

Page 13: **1.** 2, 4, 6, 40 **3.** 4, 10, 72, 18 **5.** 6, 65, 24, 130 **7.** 64 **9.** 24 **11.** 7 **13.** 3 **15.** 5 **17.** 3 **19.** $\frac{5}{7}$ **21.** $\frac{2}{5}$ **23.** $\frac{1}{3}$ **25.** $\frac{3}{4}$ **27.** $\frac{1}{4}$ **29.** $\frac{7}{10}$ **31.** $\frac{7}{9}$ **33.** $\frac{1}{3}$ **35.** $\frac{6}{7}$ **37.** $\frac{9}{14}$

Page 14: **1.** 1,2,7,14 **3.** 1,5,7,35 **5.** 1,2,4,8 **7.** 1,3,5,15 **9.** 1,2,3,5,6,10,15,30 **11.** 1,2,3,4,5,6,10,12,15,20,30,60 **13.** 1,2,3,4,5,6,8,10,12,15,20,24,30,40,60,120 **15.** 5 + 7 **17.** prime numbers: 101,103,107,109,113,127,131,137,139,149,151, 157,163,167,173,179,181,191,193,197,199

Page 15: **1.** $2^3 \times 7$;

56
8 × 7
4 × 2 × 7
2 × 2 × 2 × 7

3. 11×2^2 **5.** $3^2 \times 2$ **7.** $2^3 \times 3 \times 5$ **9.** $2^2 \times 5$ **11.** $2 \times 3^2 \times 5$ **13.** $2^4 \times 3$ **15.** 216 **17.** 99 **19.** 1512 **21.** 1617 **23.** 75 **25.** 585

Page 16: **1.** 3 **3.** 21 **5.** 28 **7.** 14 **9.** 4 **11.** 3^2; $2^2 \times 3$; 3 **13.** 2^3; $2^2 \times 5$; 4 **15.** 24 **17.** 18 **19.** 30 **21.** 40 **23.** 45 **25.** 2×3; 2^3; 24 **27.** 2^3; $2^2 \times 3$; $2^3 \times 5$; 120

Page 17: **1.** 0.08 **3.** 0.95 **5.** 2.5 **7.** 6.2 **9.** 0.4 **11.** 0.625 **13.** 0.3125 **15.** 0.16 **17.** $0.\overline{7}$ **19.** 0.75 **21.** 2.125 **23.** $1.\overline{3}$ **25.** $9.1\overline{3}$ **27.** $7.1\overline{6}$ **29.** $2.\overline{1}$ **31.** $6.\overline{4}$ **33.** 2.875 **35.** 10.3

Page 18: **1.** < **3.** < **5.** < **7.** > **9.** < **11.** < **13.** > **15.** > **17.** > **19.** > **21.** < **23.** = **25.** $\frac{2}{3}, \frac{4}{5}, \frac{6}{7}$ **27.** $\frac{2}{3}, \frac{7}{10}, \frac{3}{4}$ **29.** $\frac{1}{5}, \frac{2}{7}, \frac{3}{10}$ **31.** $0.472 < 0.857$ **33.** $0.657 < 0.662$ **35.** $0.349 > 0.289$

Page 19: **1.** $1\frac{1}{2}$ **3.** $\frac{23}{33}$ **5.** $1\frac{7}{40}$ **7.** $1\frac{11}{24}$ **9.** 17 **11.** $8\frac{5}{8}$ **13.** $17\frac{11}{36}$ **15.** $19\frac{19}{30}$ **17.** $\frac{1}{3}$ **19.** $\frac{1}{4}$ **21.** $12\frac{1}{2}$ **23.** $12\frac{1}{8}$ **25.** $12\frac{1}{21}$ **27.** $4\frac{7}{18}$ **29.** $13\frac{15}{16}$ lb **31.** $\frac{7}{24}$ yd

Page 20: **1.** $\frac{8}{15}$ **3.** $\frac{3}{35}$ **5.** $1\frac{1}{3}$ **7.** $12\frac{1}{4}$ **9.** $4\frac{1}{12}$ **11.** $23\frac{1}{3}$ **13.** $26\frac{3}{5}$ **15.** 36 **17.** $5\frac{1}{9}$ **19.** $6\frac{7}{15}$ **21.** $82\frac{2}{3}$ **23.** 30 **25.** 20 yd

Page 21: **1.** $\frac{4}{3}$ **3.** $\frac{9}{1}$ **5.** $\frac{9}{4}$ **7.** $\frac{6}{13}$ **9.** $\frac{11}{7}$ **11.** $\frac{3}{25}$ **13.** $\frac{1}{18}$ **15.** $\frac{5}{27}$ **17.** $\frac{9}{10}$ **19.** $\frac{2}{3}$ **21.** $\frac{4}{11}$ **23.** $\frac{1}{20}$ **25.** $2\frac{2}{9}$ **27.** 57 **29.** $3\frac{3}{4}$ **31.** $2\frac{4}{9}$ **33.** $16\frac{1}{2}$ **35.** $1\frac{1}{18}$ **37.** $12\frac{2}{9}$ miles

Page 22: **1.** 11 **3.** 10 **5.** 4 **7.** 6 **9.** 14 **11.** 12 **13.** 56 **15.** 2 **17.** $\frac{2}{3}$ **19.** $\frac{1}{6}$ **21.** $\frac{6}{7}$ **23.** $4\frac{1}{3}$ **25.** $6\frac{4}{21}$ **27.** $9\frac{3}{32}$ **29.** $14\frac{1}{3}$ **31.** $48\frac{32}{45}$ **33.** $27\frac{3}{5}$ **35.** about 50 in.

Page 23: **1.** 10 **3.** 8 **5.** 20 **7.** 19 **9.** 8 **11.** 4 **13.** 35 **15.** 24 **17.** 6 **19.** 7 **21.** 11 **23.** 87 **25.** 14

Page 24: **1.** $n - 8$ **3.** $5n$ **5.** $n - 9$ **7.** $12n$ **9.** $\frac{n}{4}$ **11.** $\frac{n+1}{3}$ or $\frac{n}{3} + 1$ **13.** $7n + 9$ **15.** $5n$

17. $\frac{1}{2}n$ or $\frac{n}{2}$ **19.** $5n - 4$

Page 25: **1.** $n + 4 = 10$; equation **3.** $\frac{1}{4}n - 5 > 1$; inequality **5.** $5n < 10$; inequality **7.** $3 + n < 6$; inequality **9.** $\frac{1}{2}n - 5 \geq 6$; inequality
11. $A + 6 = 14$

Page 26: **1.** 40 **3.** 23 **5.** 44 **7.** 5 **9.** 2
11. 16 **13.** 8 **15.** 80 **17.** 125 **19.** 31 **21.** 25
23. 5 **25.** 39 **27.** 50 **29.** 38 **31.** 0 **33.** 112
35. 220 **37.** 250

Page 27: **1.** 3 **3.** 14 **5.** 7 **7.** 10 **9.** 9
11. 13 **13.** 8 **15.** 7 **17.** 20 **19.** 27
21. 50 **23.** 100 **25.** 12 **27.** 84 **29.** 17
31. 91 **33.** 60 **35.** 15

Page 28: **1.** 6 **3.** 20 **5.** 20 **7.** 72 **9.** 15
11. 16 **13.** 72 **15.** 84 **17.** 98 **19.** 81
21. 184 **23.** 125 **25.** 4 **27.** 60 **29.** 65
31. 36 **33.** 192

Page 29: **1.** 7 **3.** 72 **5.** 4 **7.** 72 **9.** 7 **11.** 7
13. 4 **15.** 65 **17.** 6 **19.** 5 **21.** 16 **23.** 5
25. 15 microscopes

Page 30: **1.** 27 **3.** 100 **5.** 8 **7.** 9 **9.** 5 **11.** 6
13. 5 **15.** 8 **17.** 373 **19.** 140; 28

Page 31: **1.** $a = 5$ **3.** $y = 1$ **5.** $s = 2$
7. $x = 16$ **9.** $s = 13$ **11.** $m = 17$ **13.** $a = 102$
15. $n = 16$ **17.** $n = 11$ **19.** $d = 14\frac{3}{5}$
21. $d = 83$ **23.** $r = 5$ **25.** $8(n - 9) = 96$; $n = 21$
27. $3(n - 5) = 36$; $n = 17$

Page 32: **1.** {4} **3.** {10,11,12,...,25}
5. {25,30,35,...,50} **7.** {17,21,23,...,33} **9.** {0,1,2,3}
11. {20,10} **13.** $<$ **15.** $\geq$ **17.** $\neq$
19. {21,22,23,...,29} **21.** {0,3,6,9,12,15,18}

Page 33: **1.** Area of a Square **3.** Circumference of a Circle **5.** Area of a Triangle **7.** $S = 16$ units
9. $\ell = 5'$ **11.** $S = 18$ yd **13.** R of $T = T \div MP$
15. $D = LP \times R$ of D **17.** $68 \approx (2) \times (3.14) \times r$; $r = 10\frac{9}{11}$ ft or 10.828 ft
19. $V = 20 \times 4 \times 3$; $V = 240$ cu ft

Page 34: **1.** $^{-}5$ **3.** $^{-}4$ **5.** $^{-}17$ **7.** $>$ **9.** $<$
11. $<$ **13.** $>$ **15.** $^{-}20, ^{-}15, ^{+}2, ^{+}5, ^{+}20$
17. $^{+}15, ^{+}5, 0, ^{-}5, ^{-}19$ **19.** $^{-}$\$31.00 **21.** $^{+}500$
23. $^{-}3°C$ **25.** 25;609;14

Page 35: **1.** $^{+}9$ **3.** $^{+}7$ **5.** $^{-}5$ **7.** $^{+}10$ **9.** 0
11. $^{-}2$ **13.** $^{+}7$ **15.** $^{-}2$ **17.** $^{+}14$ **19.** $^{-}25$
21. $^{+}37$ **23.** $^{+}20$ **25.** 0 **27.** $^{-}33$ **29.** $^{-}22$
31. $^{-}18$ **33.** $^{-}2$ **35.** $^{+}11$ **37.** $^{+}15$ **39.** $^{-}1$
41. $^{-}3$ **43.** $^{-}3$ **45.** \$84 **47.** $^{-}50$

Page 36: **1.** $^{+}72$ **3.** 0 **5.** $^{+}4$ **7.** $^{+}28$ **9.** $^{+}90$
11. $^{-}93$ **13.** $^{+}10$ **15.** $^{-}3$ **17.** $^{-}9$ **19.** $^{-}9$
21. $^{-}17$ **23.** $^{+}15$ **25.** $t = ^{-}12$ **27.** $a = 0$
29. $s = ^{+}36$ **31.** $^{-}6$ **33.** $^{+}32$

Page 37: **1.** $q + 5 + (5 + 1) = 14$; $q = 3$; $n = 5$; $d = 6$; \$1.60 **3.** $8 + 4n = 40$; $n = 8$ **5.** $6n + 5 = 53$; $n = 8$ **7.** $120x = 1500$; $x = 12\frac{1}{2}$ or 12.5 hr
9. $4w = 34$; $w = 8.5$;
$A = 34 \times 8.5$; $A = 289$ cm^2

Page 38: **1.** $^{+}0.2$ **3.** $^{+}\frac{5}{3}$ **5.** $^{+}8.1$ **7.** 0 **9.** $^{+}\frac{9}{2}$
11. $>$ **13.** $<$ **15.** $>$ **17.** $<$ **19.** $<$ **21.** $<$
23. $^{-}\frac{7}{2}, ^{-}3, ^{-}0.1, ^{+}2\frac{1}{2}, ^{+}3\frac{2}{3}$
25. $^{-}1.3, ^{-}\frac{1}{3}, ^{-}0.31, ^{+}0.31, ^{+}3.10$ **27.** $^{-}5.03$
29. $^{-}4\frac{2}{3}$

Page 39: **1.** Commutative M **3.** Inverse A
5. Commutative A **7.** Identity A **9.** $^{-}12$ **11.** $^{+}1$
13. $^{-}19$ **15.** 0 **17.** $^{-}92$ **19.** $^{+}9$ **21.** $^{-}3.4$

Page 40: **1.** $^{+}8.1$ **3.** $^{+}7.5$ **5.** $^{+}15.74$ **7.** $^{+}5.59$
9. $^{-}1.95$ **11.** $^{-}5$ **13.** $^{-}0.9$ **15.** $^{-}0.07$ **17.** $^{+}2.6$
19. $^{-}0.8$ **21.** $^{+}\frac{1}{24}$ **23.** $^{-}2.77$ **25.** $^{-}7.8$
27. $^{+}0.7$ **29.** 0 **31.** $^{-}5.8$ **33.** $^{+}0.7$ **35.** $^{+}2°C$

Page 41: **1.** $^{-}1.5$ **3.** $^{-}3$ **5.** $^{-}2.4$ **7.** $^{+}21.6$
9. $^{-}20.4$ **11.** 0 **13.** $^{+}7\frac{1}{2}$ **15.** $^{-}5$ **17.** $^{+}1\frac{1}{2}$
19. $^{+}2.6$ **21.** $^{-}9.5$ **23.** $^{-}3.01$ **25.** $^{-}5.74$
27. $^{-}$\$6.94

Page 42: **1.** $^{+}4$ **3.** $^{+}5$ **5.** $^{+}1$ **7.** $^{+}5$
9. $^{-}14$ **11.** $^{+}53$ **13.** $^{+}8$ **15.** $^{+}19$ **17.** $^{+}1.3$
19. $^{+}3$ **21.** $^{-}26$ **23.** $n + 7 = ^{-}12, n = ^{-}19$
25. $n + 15 = 41; n = ^{+}26$

Page 43: **1.** $^{+}6$ **3.** $^{+}36$ **5.** $^{+}16$ **7.** $^{+}28$ **9.** $^{-}12$
11. $^{-}20$ **13.** $^{+}63$ **15.** $^{-}27$ **17.** $^{+}0.024$
19. $^{-}10.28$ **21.** $^{-}0.0607$ **23.** $^{+}\frac{3}{8}$ **25.** $^{-}\frac{3}{16}$
27. $^{-}\frac{5}{27}$ **29.** $^{-}1\frac{7}{8}$ **31.** $^{-}20$ **33.** $^{-}20$ **35.** $^{-}35$
37. $^{-}18$ **39.** $^{-}200$

Page 44: **1.** $^{+}4; ^{+}4$ **3.** $^{+}63; ^{+}63$ **5.** $^{+}3; ^{+}3$ **7.** $^{-}8$
9. $^{-}4$ **11.** $^{+}9$ **13.** $^{+}8$ **15.** $^{-}9$ **17.** $^{+}2$ **19.** $^{-}1$
21. $^{-}0.7$ **23.** $^{+}11\frac{3}{5}$ **25.** $^{-}3\frac{1}{2}$ **27.** $^{+}1\frac{1}{2}$ **29.** $^{-}9$

Page 45: **1.** $4n = ^{-}55; n = ^{-}13\frac{3}{4}$ or $^{-}13.75$
3. $^{-}9n = ^{-}36; n = ^{+}4$ **5.** $^{+}7$ **7.** $^{+}9$ **9.** $^{-}0.08$
11. $^{-}35.01$ **13.** $^{-}\frac{1}{63}$ **15.** $^{-}0.9$ **17.** $^{-}1600$
19. $^{-}\frac{3}{4}$ **21.** $^{-}72$ **23.** $^{-}288$ **25.** $^{-}12.5$ **27.** $^{+}5.6$
29. $^{+}13$ **31.** $^{+}192$

Page 46: **1.** $LCM = 10; x = ^{-}5$
3. $LCM = 10; x = 2.4$ **5.** $LCM = 10; y = 120$
7. $LCM = 3; s = 90$ **9.** $y = 12$ **11.** $y = 3.5$
13. $c = 3\frac{1}{8}$ **15.** $r = ^{-}5$ **17.** $y = 3\frac{1}{10}$
19. $x = 3.6$ **21.** $t = 11$ **23.** $n = 8$ **25.** $x = 3.6$

Page 47: **1.** 2*C*, 1*A*, 3*B* **3.** Maria: \$108,000; Megan: \$72,000; Ramon: \$144,000; Jeff: \$108,000
5. Moya: 15; Catherine: 15

Page 48: **7.** \$17 **9.** Mighty Mo: \$201.20; Tiny Tim: \$253.24; Sweet Pea: \$239.36

Page 49: **1.** 1000 **3.** 10,000 **5.** 10,000,000,000
7. 10^5 **9.** 10^0 **11.** 10^7 **13.** 84 **15.** 948
17. 9300 **19.** 384.1 **21.** 643,110 **23.** 0.0793
25. 0.5063 **27.** 0.0000106 **29.** 0.0000042
31. 0.007009 **33.** 8.22×10^7; 6321.004×10^3;

5081.426×10^3; 4.321×10^6; 42.21021×10^5
35. $84 \div 10^2$; $8400 \div 10^5$; $8.4 \div 10^3$; $8400 \div 10^7$; $0.84 \div 10^6$

Page 50: **1.** 843.074 **3.** 1204.56 **5.** 0 hundredth **7.** 2 ten thousandths **9.** 1 hundred thousand **11.** 8 hundredths **13.** (9×10^{-2}) **15.** (1×10^{-4}) **17.** $(6 \times 10^1) + (6 \times 10^0) + (3 \times 10^{-1})$ **19.** $(8 \times 10^2) + (3 \times 10^1) + (8 \times 10^0) + (5 \times 10^{-4}) + (2 \times 10^{-5})$ **21.** $(4 \times 10^0) + (8 \times 10^{-4}) \times (4 \times 10^{-5}) + (6 \times 10^{-6})$

Page 51: **1.** 2 **3.** $^-5$ **5.** 6.2×10^4 **7.** 6×10^{-6} **9.** 25,200,000,000 **11.** 202,500,000,000 **13.** 0.5985 **15.** 10^3 **17.** 10^3 **19.** 6^4 **21.** 8^1 **23.** 2×10^1 **25.** 1.24×10^{-3} **27.** 1.07×10^8 km **29.** 637,000,000 or 6.37×10^8 m^2

Page 52: **1.** 168; 2514; 217,251 **3.** 948; 66; 27,684; 186,630 **5.** 324; 81,648; 32,166; 386,451 **7.** 247,116; 920,424; 327,164; 367,148 **9.** 112,926; 617,122; 122,221 **11.** No; Yes; No **13.** No; Yes; No **15.** No; Yes; No **17.** Yes; No; Yes **19.** Yes; No; Yes **21.** Yes; Yes; Yes **23.** 18 **25.** Yes; 86

Page 53: **1.** 24; 29; 34 **3.** 46; 44; 42 **5.** 16; 32; 64 **7.** 36; 43; 50 **9.** 12; 12.5; 13 **11.** $\frac{5}{3}$; 2; $\frac{7}{3}$ **13.** 11; 16; 22 **15.** 13.1; 13.3; 13.5 **17.** 17, 18, 23 **19.** 56; 50; 45 **21.** 19; 22; 25 **23.** 81; 243; 729 **25.** 2.2; 2.5; 2.8 **27.** 82; 77; 71 **29.** 10; 11.5; 13 **31.** 12; 6; 9 **33.** 32; 128; 64

Page 54: **1.** $p \wedge q$ **3.** $q \rightarrow r$ **5.** $r \rightarrow (p \wedge q)$
7. Love is everything and silence gives consent.
9. If silence gives consent, then time is a thief.
11. Time is a thief or love is everything.

13.

p	q	$\sim p$	$\sim p \rightarrow q$
T	T	F	F
T	F	F	T
F	T	T	T
F	F	T	F

15.

p	q	$\sim p$	$\sim p \wedge q$
T	T	F	F
T	F	F	F
F	T	T	T
F	F	T	F

17.

p	q	$p \vee q$	$\sim(p \vee q)$
T	T	T	F
T	F	T	F
F	T	T	F
F	F	F	T

Page 55: **1. Converse:** If a number is zero, then it is not divisible by one.
Inverse: If a number is divisible by one, then it is not zero.
Contrapositive: If a number is not zero, then it is divisible by one.
3. Converse: If Stretch is a snake, then Stretch is a python.
Inverse: If Stretch is not a python, then Stretch is not a snake.
Contrapositive: If Stretch is not a snake, then Stretch is not a python.
5. T, F, F, T **7.** F, T, T, F **9.** T, T, T, T

Page 56: **1.** $\frac{9}{10}$ **3.** $\frac{4}{9}$ **5.** $\frac{1}{3}$ **7.** $\frac{3}{5}$ **9.** $\frac{\$450}{1}$ **11.** $\frac{1}{\$6.25}$ **13.** $\frac{25}{2}$ **15.** $\frac{6}{1}$ **17.** $\frac{1}{4}$ **19.** $\frac{5}{9}$ **21.** $\frac{2}{1}$ **23.** $\frac{1}{2}$ **25.** $\frac{6}{5}$ **27.** $\frac{8}{17}$ **29.** $\frac{6}{5}$ **31.** $\frac{4}{1}$ **33.** $\frac{4}{1}$ **35.** $\frac{9}{1}$ **37.** 27:43 or $\frac{27}{43}$

Page 57: **1.** = **3.** ≠ **5.** = **7.** ≠ **9.** = **11.** ≠ **13.** 3 **15.** 6 **17.** 7 **19.** $1\frac{1}{2}$ **21.** 4 **23.** 0.4 **25.** $1\frac{1}{4}$ **27.** 5 **29.** 0.4 **31.** 19.2 km

Page 58: **1.** $1.25 **3.** 215 plates **5.** $1.26 **7.** 30¢ **9.** 520 Kilometers **11.** $2365

Page 59: **1.** 10 km **3.** 30 cm **5.** 16 cm **7.** 44 cm **9.** 2.1 cm **11.** 0.075 cm **13.** 12.24 cm

Page 60: **1.** 2 **3.** 2 **5.** 10 **7.** 6 people **9.** $10\frac{1}{2}$ hr **11.** 1 hr **13.** 4 hr

Page 61: **1.** 16; 48 **3.** 60; 90; 120 **5.** 120; 150; 180 **7.** 15; 60; 90 **9.** John: $45, Martin: $30 **11.** 312 birthday cards; 156 anniversary cards; 52 get-well cards; **13.** Ms. Wayne: 1680 votes; Mr. Edwards: 960 votes.

Page 62: **1.** $x = 6.8$; $y = 5.1$ **3.** $\overline{HK}$ **5.** 73° **7.** 0.75; 37° **9.** 0.8; 37° **11.** 0.6; 53° **13.** 37.6 **15.** 82.41

Page 63: **1.** 0.3; 30% **3.** 25: 100; 0.25; 25% **5.** 80: 100: $\frac{4}{5}$; 80% **7.** 42: 100; $\frac{21}{50}$; 42% **9.** $\frac{5}{8}$; 0.625; 62.5% **11.** 6: 10; $\frac{3}{5}$; 60% **13.** 12: 5: 100; 0.125; $12\frac{1}{2}$% **15.** 2.5 **17.** 10: 100 **19.** 4.5 **21.** 37:100 **23.** $\frac{2}{50}$ **25.** 0.15 **27.** 87:100

Page 64: **1.** $\frac{2}{5}$ **3.** $\frac{8}{25}$ **5.** $\frac{1}{8}$ **7.** $\frac{1}{6}$ **9.** $\frac{1}{7}$ **11.** $\frac{5}{8}$ **13.** $1\frac{1}{4}$ **15.** $2\frac{3}{4}$ **17.** $\frac{9}{200}$ **19.** $\frac{7}{200}$ **21.** $\frac{1}{250}$ **23.** $1\frac{1}{20}$ **25.** 0.65 **27.** 0.98 **29.** 0.24 **31.** 0.04 **33.** 0.355 **35.** 0.751 **37.** 0.0225 **39.** 0.074 **41.** 1 **43.** 2.5 **45.** 0.006 **47.** 0.004 **49.** $83\frac{1}{3}$%

Page 65: **1.** 4 **3.** 21 **5.** 3.5 **7.** 8 **9.** 28.08 **11.** 0.34 **13.** 7 **15.** 41 **17.** 28 **19.** 20.15 **21.** 0.225 **23.** 1404 cars **25.** 15 students

Page 66: **1.** 25% **3.** 20% **5.** 24% **7.** $66\frac{2}{3}$% **9.** $11\frac{1}{9}$% **11.** 350% **13.** $8\frac{1}{3}$% **15.** 225% **17.** 40% **19.** $66\frac{2}{3}$%

Page 67: **1.** 1400 **3.** 32 **5.** 15 **7.** 24 **9.** 51 **11.** 52 **13.** 264 **15.** 500 **17.** 5 **19.** 6 **21.** $14\frac{2}{7}$% **23.** 54 **25.** 24

Page 68: **1.** $\frac{1}{3}$ **3.** $\frac{1}{8}$ **5.** $\frac{1}{3}$ **7.** $\frac{2}{5}$ **9.** $\frac{1}{5}$ **11.** 7 **13.** 16 **15.** 27 **17.** $33\frac{1}{3}$; b **19.** 60; b **21.** 112 **23.** 147 **25.** 480

Page 69: **1.** 5; 11.1% **3.** 8; 13.3% **5.** 7; 12.1% **7.** 6; 8.1% **9.** 6; 10.7% **11.** 36.4% **13.** 33.3% **15.** 6.7% increase

Page 70: **1.** *b* **3.** *c* **5.** *c* **7.** *c* **9.** *d* **11.** *c* **13.** *c* **15.** *d*

Page 71: **17.** = **19.** 6.675 **21.** $^{-}7.4$ **23.** $n + 15 = 41$; 26 **25.** $4n + 5 = 17$; $n = 3$ **27.** 555 **29.** If it is spring, then it is raining.

Page 72: **1.** $10 **3.** $65.33 **5.** $20.13 **7.** $6.91 **9.** $33.83 **11.** $17.04 **13.** $99.40 **15.** $315.93 **17.** $729.01 **19.** $15,901.80

Page 73: **1.** $52.33; $104.66 **3.** $5.12; $26.88 **5.** $4.48; $51.52 **7.** $1.05; $13.90 **9.** $12.35; $64.85 **11.** $20; $60 **13.** $37.25, $149 **15.** $210; $1190

Page 74: **1.** 30% **3.** 14% **5.** $66\frac{2}{3}$% **7.** $4500 **9.** $63 **11.** $85 **13.** 5%

Page 75: **1.** $4.47 **3.** $10.54 **5.** $6.08 **7.** $9.69 **9.** $2.80 **11.** $17.43; $366.03 **13.** $4.98; $104.48 **15.** $13.99;$293.84 **17.** $11.40

Page 76: **1.** $120 **3.** $90 **5.** $261.90 **7.** $141.75 **9.** $19.60 **11.** $38.94 **13.** $367.62 **15.** $344

Page 77: **1.** 8% **3.** 15% **5.** 5% **7.** $225 **9.** $4050 **11.** $4550 **13.** $5000 **15.** 10%

Page 78: **1.** $480 **3.** $63 **5.** $307.20 **7.** $22.50 **9.** $113.40 **11.** $2194.80 **13.** $547.20

Page 79: **1.** $I = \$300$; $P = \$5300$; $I = \$318$ **3.** Interest: $1515.75; Total: $9015.75 **5.** Mr. Mendoza ($50.31)

Page 80: **1.** $2.50 **3.** $.89 **5.** $1.95 **7.** $50 **9.** $25 **11.** $50 **13.** $12.50 **15.** $20 **17.** $183.63

Page 81: **1.** $\frac{1}{6}$ **3.** $\frac{1}{3}$ **5.** 0 **7.** $\frac{2}{3}$ **9.** $\frac{1}{6}$ **11.** $\frac{1}{8}$ **13.** $\frac{1}{8}$ **15.** 0 **17.** $\frac{1}{2}$ **19.** $\frac{1}{5}$ **21.** 0 **23.** $\frac{1}{2}$

Page 82: **1.** $\frac{1}{16}$ **3.** $\frac{1}{16}$ **5.** $\frac{1}{32}$ **7.** $\frac{1}{16}$ **9.** $\frac{1}{8}$ **11.** $\frac{1}{64}$; $\frac{1}{56}$ **13.** $\frac{1}{64}$; $\frac{1}{56}$ **15.** $\frac{1}{4}$; $\frac{2}{7}$ **17.** $\frac{1}{4}$; $\frac{3}{14}$ **19.** $\frac{1}{16}$; $\frac{1}{14}$

Page 83: **1.** 720 **3.** 40,320 **5.** 3,628,800 **7.** 40,314 **9.** 210 **11.** 24 **13.** 120, $\frac{1}{120}$

Page 84: **1.** 15; 25% **3.** 20; $33\frac{1}{3}$% **5.** 100 **7.** 150 **9.** 200 **11.** 22; 23% **13.** 31; 33% **15.** 18 **17.** 45 **19.** 69

Page 85: **1.** 600 **3.** hockey and soccer **5.** 7000 books **7.** 2500 books

Page 86: **1.** 11; 13; 8; 10; 7; 10

Page 87: **1.** 30 **3.** 94 **5.** 42 **7.** 70 **9.** 415

Page 89: **1.** 90° **3.** 120° **5.** 240° **7.** 216° **9.** $37\frac{1}{2}$; 25%; 25%; $12\frac{1}{2}$% **11.** 126; 86.4 (86); 64.8 (65); 43.2 (43); 14.4 (14); 8.28 (8); 7.56 (8); 4.68 (5); 4.68 (5)

Page 90: **1.** range = 72; median = $51; mean = $54.90; mode = none **3.** range = 36; median = 45; mean = 45.875; mode = none **5.** average = 0.313; range = 0.092; median = 0.308 **7.** mean = 9.34; mode = 5.6

Page 91: **1.** *a* **3.** *b* **5.** *c* **7.** *e* **15.** True **17.** False **19.** True **21.** False

Page 92: **1.** 50; acute **3.** 120; obtuse **9.** 2° **11.** 77° **13.** 166° **15.** 88° **17.** *a*

Page 94: **1.** $\overleftrightarrow{SW} \parallel \overleftrightarrow{TY}$ **3.** 90° **5.** 90° **7.** $\angle 1 - \angle 2$; $\angle 2 - \angle 4$; $\angle 3 - \angle 1$; $\angle 4 - \angle 3$; $\angle 5 - \angle 7$; $\angle 6 - \angle 5$; $\angle 7 - \angle 8$; $\angle 8 - \angle 6$ **9.** $\angle 2 - \angle 6$; $\angle 4 - \angle 8$; $\angle 1 - \angle 5$; $\angle 3 - \angle 7$ **11.** $\overleftrightarrow{AB} \parallel \overleftrightarrow{CD}$ **13.** $\angle CFH$ **15.** $\angle AEH$

Page 95: **1.** 70°; acute isosceles **3.** 61°; acute scalene **5.** 88; acute scalene **7.** 72°; acute scalene **9.** 65°; obtuse scalene **11.** 60°; acute equilateral **13.** 45°; right isosceles **15.** isosceles **17.** obtuse

Page 96: **1.** $\overline{XZ}$ **3.** $\overline{YZ}$ **5.** $\angle M$ **7.** ASA **9.** SAS

Page 97: **1.** 3 **3.** $1\frac{1}{2}$ **5.** 14 **7.** 6000 **9.** 7040 **11.** 10 lb **13.** 2 gal 2 qt **15.** 8 gal 2 qt **17.** 9 qt **19.** $69.36

Page 98: **1.** cm **3.** m **5.** cm **7.** 8.4 **9.** 0.0807 **11.** 36 **13.** 0.24 **15.** 24 **17.** 24 **19.** 400 **21.** 0.0005 **23.** 98 100 **25.** 0.026 **27.** 0.03 **29.** 0.091 **31.** 0.072 **33.** 600 000 **35.** 8100 **37.** 826 **39.** 16.047 **41.** 3.002 **43.** 80 005 **45.** 6019 **47.** 3.46

Page 99: **1.** 0.5; 500 **3.** 0.406; 40.6; 406; 40 600 **5.** 68; 6800; 68 000; 6 800 000 **7.** 1500 **9.** 25 000 **11.** 620 **13.** 0.0046 **15.** 0.3 **17.** = **19.** = **21.** > **23.** Principal's office to library; 0.14 m

Page 100: **1.** 2.5 cm **3.** 48 mm **5.** 6.1 dm **7.** 7314 m **9.** 142 cm **11.** 0.1 cm; 0.05 cm; 16.1 ± 0.05 cm **13.** 1 dm; 0.5 dm; 7 ± 0.5 dm **15.** 0.1 m; 0.05 m; 2.7 ± 0.05 m **17.** 0.1 dam; 0.05 dam; 9.6 ± 0.05 dam **19.** 1 cm; 0.5 cm; 18 ± 0.5 cm

Page 101: **1.** 1 mi **3.** Town A: 62%; Town B: 39%; Both: 50% **5.** 16 students

Page 102: **7.** square 12 **9.** 90°; 22°; 68°

Page 103: **1.** 44.8 cm **3.** 25.41 m **5.** 30 m **7.** 25.2 cm **9.** 170.4 cm **11.** about 29 yd

Page 104: **1.** 240 cm^2 **3.** $69\frac{4}{9}$ in.^{2} **5.** 26.46 m^2 **7.** $44\frac{1}{3}$ ft^2 **9.** 213 m^2 **11.** 0.72 m^2 or 72 dm^2 **13.** 20 yd^2 or 180 ft^2 **15.** 20 containers **17.** 0.96 m^2 **19.** square: 144 ft^2; rectangle: 140 ft^2; square

Page 105: **1.** 104 cm^2 **3.** 79.2 cm^2 **5.** 6 m^2 **7.** 8.4 dm^2 **9.** 2340 cm^2 **11.** 47.5 yd^2 **13.** 7.5 m^2 **15.** 165 ft^2

Page 106: **1.** 36 cm^2 **3.** 32.76 m^2 **5.** $1\frac{1}{9}$ **7.** 1734 cm^2 **9.** 2200 cm^2 **11.** 76.5 ft^2 or $8\frac{1}{2}$ yd^2

Page 107: **1.** 9.42 m **3.** 94.2 mm **5.** 5.024 km **7.** 22 cm **9.** 132 dm **11.** 17.6 m **13.** 66 cm **15.** 28.26 yd **17.** 13.2 m

Page 108: **1.** 154 **3.** 452.16 **5.** $218\frac{16}{63}$ or 218.06 **7.** 12.56 **9.** 615.44 mm^2 **11.** 113.04 in.^{2} **13.** 113.04 in.^{2} **15.** 38.6851 m^2 **17.** 0.2197 cm^2 **19.** 142 458.66 km^2 **21.** 7234.56 m^2 **23.** 16 times **25.** 29.7083 m^2

Page 109: **1.** 27.125 m^2 **3.** 625 cm^2 **5.** 113.04 cm^2 **7.** 17.415 cm^2 **9.** $51.\overline{3}$ cm^2

Page 110: **1.** 12 m **3.** 1.5 ft **5.** 0.9 m **7.** $6\frac{3}{4}$ ft **9.** 72.8 cm^2 **11.** $4\frac{1}{2}$ yd **13.** 300 m **15.** 324 ft^2 **17.** 14 yd **19.** 120.9 cm^2 **21.** 2 ft **23.** 3 yd **25.** 0.9 m

Page 111: **1.** $\frac{71}{1}$ **3.** $\frac{14}{5}$ **5.** $\frac{-314}{1}$ **7.** $\frac{306}{100}$ or $\frac{153}{50}$ **9.** 100; $63.\overline{63}$; $0.\overline{63}$; 63.00; $\frac{63}{99} = \frac{7}{11}$ **11.** $\frac{16}{99}$ **13.** $\frac{1}{9}$ **15.** $\frac{79}{90}$ **17.** $\frac{5}{6}$ **19.** $\frac{7}{9}$ **21.** $\frac{5}{12}$

Page 112: **1.** 0.25 **3.** 0.0036 **5.** $\frac{4}{49}$ **7.** 0.81 **9.** $\frac{36}{49}$ **11.** 0.64 **13.** 0.0081 **15.** $\frac{4}{81}$ **17.** 0.0049 **19.** 0.0025 **21.** $\frac{7}{12}$ **23.** $\frac{9}{16}$ **25.** 1.9 **27.** 1.7 **29.** 1.5 **31.** $\frac{5}{11}$ **33.** 2 and 3

41. 10 and 11 **43.** 18 and 19 **45.** 8 and 9 **47.** 13 and 14 **49.** 12 and 13

Page 113: **1.** 4; 486; 46 **3.** 4; 486; 24.6 **5.** 48 **7.** 83 **9.** 235 **11.** 301 **13.** 6.04 **15.** 0.69 **17.** 44.6 **19.** 36.2

Page 114: **1.** rational **3.** irrational **5.** rational **7.** rational **9.** rational **11.** ±4 **13.** ±8 **15.** ±7 **17.** 0.25 **19.** $\frac{11}{14}$ **21.** 0.0085

Page 115: **1.** $\{r: r > {}^-8\}$ **3.** $\{x: x \leq 5\}$ **5.** $\{b: b \geq 6\}$ **7.** $\{x: x \leq 48\}$ **9.** $\{t: t < 6\}$ **11.** $\{y: y \leq {}^-57\}$ **13.** $\{a: a < 21\}$ **15.** $\{k: k \leq 11.1\}$ **17.** $\{a: a \neq 4\}$ **19.** $\{n: n \leq 1\frac{1}{5}\}$

Page 116: **1.** 15 m **3.** 25 m **5.** 14 m **7.** 84 m **9.** 200 m **11.** 20 m

Page 117: **1.** 304 in.^{2} **3.** 115.2 m^2 **5.** 34.56 cm^2 **7.** 864 cm^2 **9.** 1944 in.^{2} **11.** 1536 ft^2 **13.** 1940 cm^2 **15.** 4500 m^2 **17.** 12 304 cm^2

Page 118: **1.** 336 in.^{2} **3.** 152 ft^2 **5.** 336 ft^2 **7.** 132 m^2 **9.** 14.7 m^2

Page 119: **1.** 1280 m^2 **3.** 62.4 cm^2 **5.** 1372 in.^{2} **7.** 1112 cm^2 **9.** 3600 cm^2

Page 120: **1.** 678.24 in.^{2} **3.** 748 ft^2 **5.** 244.92 yd^2 **7.** 1188 ft^2 **9.** 13 728 cm^2 **11.** 785 ft^2 **13.** 179.3568 m^2 **15.** 31 550.72 cm^2

Page 121: **1.** 266.9 cm^2 **3.** 5.495 m^2 **5.** 113.04 m^2 **7.** 60.7904 cm^2 **9.** 21 703.68 cm^2 **11.** 6600 cm^2 **13.** 5544 cm^2

Page 122: **1.** 216 ft^3 **3.** 1.728 m^3 **5.** 54 m^2 **7.** It is eight times greater. **9.** yes; no **11.** 5832 **13.** 0.216 m^3

Page 123: **1.** 114.39 m^3 **3.** 1092 in.^{3} **5.** 529 200 cm^3 **7.** 9484.8 cm^3 **9.** 28 800 cm^3 **11.** 1 m^3 (1.0944 m^3)

Page 124: **1.** 206.64 cm^3 **3.** 94.5 in.^{3} **5.** 906.36 m^3 **7.** 360 yd^3 **9.** 3 ft^3

Page 125: **1.** 138.6 m^3 **3.** $31\frac{2}{3}$ cm^3 **5.** 14.7 cm^3 **7.** 256 in.^{3} **9.** 36 cm^3

Page 126: **1.** 86.24 m^3 **3.** 125.6 dm^3 **5.** 462 cm^3 **7.** 8800 cm^3 **9.** 282.6 in.^{3} **11.** 69.08 in.^{3} **13.** 33 264 cm^3 **15.** cylindrical container holds 66.6 yd^3 more

Page 127: **1.** milliter **3.** kiloliter **5.** liter **7.** 8 **9.** 70 **11.** 6 **13.** 45 **15.** 30 **17.** 18 000 **9.** 3.4 **21.** 8.756 **23.** 531.441 **25.** 18.48 L **27.** 565.2 kL

Page 128: **1.** grams **3.** kilograms **5.** grams **7.** grams **9.** 6 **11.** 6 **13.** 25 **15.** 135 **17.** 2.5 **19.** 7 **21.** 1.8 kg

Page 129: **23.** e **25.** a **27.** d **29.** 5000 **31.** 6800 **33.** 4.9 **35.** 2900 **37.** 3800 **39.** 890 **41.** 11.3 **43.** 2 g **45.** 1000 kg **47.** 10.5 dm^3; 10.5 kg **49.** 35 kL; 35 000 kg **51.** 10 L; 10 kg

Page 130: **1.** $x = {}^-3$ **3.** $x = 4$ **5.** $x = {}^-3$ **7.** $x \geq {}^-2$ **9.** $x \leq {}^+2$ **11.** $x > {}^+2$ **13.** $x \leq {}^+\frac{1}{2}$ **15.** $x = {}^+2$ **17.** $x > 0$ **19.** ${}^-3 \leq x \leq {}^+2$

Page 131: **21.** *A, O, E, S* **23.** *C, T, P, G*

Page 132: **1.** 3; 4; 5; 6; 7; 8; 9 **3.** 5; 6; 7; 8; 9; 10; 11 **5.** ⁻10; ⁻7; ⁻4; ⁻1; 2; 5; 8 **7.** 1; (⁻2, 1) **9.** 0 + 3; 3; (0, 3) **11.** 2 + 3; 5; (2, 5) **13.** 3(⁻1); ⁻3; (⁻1, ⁻3) **15.** 3(1); 3; (1, 3) **17.** 12; 11; 8; 7 **19.** ⁻5; ⁻6; ⁻7; ⁻8

Page 133: **1.** 1; 2; 3; 4; 5 **3.** 7; 5; 3; 1; ⁻1

Page 134: **1.** ⁻1; 0; 1; 2; 3; 4; 5; ⁻9; ⁻6; ⁻3; 0; 3; 6; 9; (1,3) **3.** (⁻3, 5) **5.** (⁻2, ⁻1)

Page 135: **1.** $(^-2, ^-12)$ **3.** $(^-4, ^-1)$ **5.** $(2, 7)$ **7.** $(2, ^-8)$ **9.** $(1, ^-1)$ **11.** $(^-1, 10)$ **13.** $(^-2, 2)$ **15.** $(11, 15)$

Page 136: **1.** $y > 3x - 5$; dotted **3.** $y < 3 - 2x$; dotted **5.** $y \leq \frac{2}{3}x - \frac{1}{3}$; solid **7.** $y \geq \frac{-x}{2} + 3$; solid **11.** solid

Page 137: **1.** *a* **3.** *b*

Page 138: **1.** 10; 5; 2; 1; 2; 5; 10 **3.** 7; 2; $^-1$; $^-2$; $^-1$; 2; 7 **5.** $^-9$; $^-4$; $^-1$; 0; $^-1$; $^-4$; $^-9$

Page 139: **1.** *c* **3.** *a* **5.** *b* **7.** *a* **9.** *d* **11.** *a* **13.** *c* **15.** *c*

Page 140: **15.** *c* **17.** *b* **19.** *d* **21.** *d* **23.** *c* **25.** \$206.25 **27.** 90°; 37° **29.** 924 in.^{2}

Page 141: **1.** binomial; D = 4 **3.** trinomial; D = 3 **5.** monomial; D = 5 **7.** monomial; D = 0 **9.** trinomial; D = 3 **11.** trinomial; D = 6 **13.** -27 **15.** $\frac{-1}{81}$ **17.** 0 **19.** $\ell = 23$; $w = 8$; $A = 184$ **21.** height = 4; bases = 9; 5; area = 28

Page 142: **23.** edge = 10; volume = 1000 **25.** $\ell = 7$; $w = 3$; $h = 4$; $V = 84$ **27.** $r = 7$; $h = 17$; $V = 2618$ **29.** *c* **31.** *d*

Page 143: **1.** $35x^2y^3$ **3.** $6c - 8d$ **5.** $\frac{5}{8}pq - 1\frac{1}{9}cd$ **7.** $b^2 - 4.8n^2$ **9.** $-5a^2 + 28a - 1$ **11.** $63m^2 - 36mn + 2n^2$ **13.** $17d^2 - 9$ **15.** $74a^2 - 86b^2$ **17.** $35x^2y + 70xy^2 - 13xy$ **19.** y^2

Page 144: **21.** $20x - 8$ **23.** $24x - 2$ **25.** $10 + a - 3a^2$ **27.** $10 + 4ab - 7a^2b^2 + 9a^4b^3$ **29.** $-15x^5 - x^4y + x^3y^2 + x^2y + 3$ **31.** $5a^2 + 2b - 6$ **33.** $4x^2 + 11x - 31$

Page 145: **1.** a^{11} **3.** n^7 **5.** x^5 **7.** 1 **9.** m^{15} **11.** d^{36} **13.** x^7y^7 **15.** $81x^8y^{12}$ **17.** $1000a^{11}6b^{12}$ **19.** $a^8b^{10}c^{12}$ **21.** $x^{14}y^6z^{10}$ **23.** $32x^{20}y^5z^5$ **25.** 81 **27.** $\frac{1}{16}$ **29.** 49

Page 146: **1.** $60x^{11}$ **3.** $-206a^6$ **5.** $64x^9y^{12}$ **7.** $-80c^{12}$ **9.** $\frac{8}{27}a^{15}b^{12}c^6$ **11.** $-m^5n^5p^5$ **13.** $15a^3b^3 + 20a^5b^4 + 30a^6b^2$ **15.** $-63x^5y^7 + 42x^7y^5 - 49x^9y^3$ **17.** $-2m^6n^8 + 3m^7n^9 - 7m^5n^7$ **19.** $-2.38x^5y^3 - 3.06x^2y^5$ **21.** $8a^4b^3c^5$ **23.** $3a^2k$ **25.** $280xy$ cents **27.** $35a^5b^2$ cm^2

Page 147: **1.** $x^2 + 7x + 12$ **3.** $x^2 - 13x + 42$ **5.** $a^2 + 8a - 65$ **7.** $c^2 + 5c - 84$ **9.** $m^2 + 13m + 45$ **11.** $a^2 - 100$ **13.** $c^2 + 24c + 144$ **15.** $x^2 - 22x + 121$ **17.** $6x^2 + 19x + 15$ **19.** $6x^2 - 35x + 36$ **21.** $24x^2 + 47x - 21$ **23.** $4x^2 - 21x - 18$ **25.** $36x^2 - 121$ **27.** $49x^2 + 70x + 25$ **29.** $100x^2 - 20x + 1$ **31.** $77a^2 - ab - 6b^2$ **33.** $169a^2 - 25b^2$

Page 148: **35.** $48a^2 + 34ab - 105b^2$ **37.** $m^2 - 9n^2$ **39.** $121a^2 - 132ab + 36b^2$ **41.** $c^2 - 28cd + 196d^2$ **43.** $a^3 - 125$ **45.** $x^3 + 6x^2 + 12x + 8$ **47.** $a^3 + 3a^2 - 17a + 6$ **49.** $6 + a - 6a^2 - a^3$ **51.** $8x + 15$ **53.** $x^3 - 11x$ **55.** $(3x^2 + 20x - 63)$in.^{2} **57.** $(121x^2 - 44x + 4)$ cm^2

Page 149: **1.** $3(2x + 3y)$ **3.** $a^2b^2c^3(ab - bc + ac)$ **5.** $5(r^6 + 2a^2)$ **7.** $5a^2(5a^4 + 2)$ **9.** $25x^2(5x^3 + 7x^2 + 8)$ **11.** $p^2q^3(pq - r)$ **13.** $12x^3y^2z^2(3x^3z^2 + 4x^2)$ **15.** $p^8(1 + p^4 + p^{12})$ **17.** $13a^4b^3(ab^2 - 13)$ **19.** $3(49 - 12x^5 + 96x^7)$ **21.** $9(21x^9 + 30x^4 + 16)$ **23.** $13ab(2a^2b + 3ac + 5bc^2)$ **25.** $8x^2y^3(4x^3 + 5xy + 6y^2)$

Page 150: **27.** $10cd2(4c3 - 5cd + d)$ **29.** $5x^3y(-2x^2 - 15xy - 9y^2)$ **31.** $11a^4bc^2(-3ab^2 - 4bc - 2ac^2)$ **33.** $0.2x^2y^2z^2(xy^3 - 4y^2z + 6x^2z^2)$ **35.** $0.3a^2b^2c(9ab + bc - 2a^2c)$ **37.** $2xy(x - y)$ **39.** $c^4 + 24$

Page 151: **1.** $(x + 5)(x + 1)$ **3.** $(y - 19)(y - 1)$ **5.** $(b - 16)(b - 4)$ **7.** $(y - 3)(y - 8)$ **9.** $(a + 16)(a + 1)$ **11.** $(x - 18)(x + 3)$ **13.** $(a - 5)(a + 2)$ **15.** $(x - 36)(x - 5)$ **17.** $(y + 10)(y + 12)$ **19.** $(m - 9)(m + 4)$ **21.** $(c - 4d)(c - 11d)$ **23.** $(y - 18)(y - 8)$ **25.** $(n - 9)(n - 1)$ **27.** $(m^2 - 7n^2)(m^2 + 15n^2)$ **29.** $(x + 15)(x + 8)$ **31.** $(d + 13)(d + 5)$ **33.** $(c^2 - 23d^2)(c^2 + d^2)$ **35.** $(x - 12)(x - 5)$

Page 152: **37.** $(a + 4b)(a + 2b)$ **39.** $(x - 4y)(x - 8y)$ **41.** $(b - 10c)(b + 2c)$ **43.** $(y + 16z)(y - 3z)$ **45.** $(m + 7n)(m + 2n)$ **47.** $(r - 5s)(r - 10s)$ **49.** $(a + 7b)(a + 9b)$ **51.** $(x - 4)(x - 3)$ **53.** $(x + 14)(x + 3)$

Page 153: **1.** $(3x + 1)(x + 1)$ **3.** $(12 - 7x)(5 + 2x)$ **5.** $(6x + 5)(x - 2)$ **7.** $(5a - 2)(3a - 1)$ **9.** $(4r + 3s)(6r + 5s)$ **11.** $(6a - 1)(5a - 2)$ **13.** $(7x + 2)(5x + 3)$ **15.** $(7a + 4b)(3a - 10b)$ **17.** $(3x - 4)(3x + 1)$ **19.** $(2x + 1)(8x - 1)$ **21.** $(8c - 9d)(3c + d)$ **23.** $(10x - 1)(2x - 3)$ **25.** $(5x - 1)(x + 1)$ **27.** $(9a - 5b)(5a + 9b)$ **29.** $(6x - 7)(2x - 1)$ **31.** $(5x - 2)(3x - 4)$ **33.** $(3x^3 + 11y)(2x^3 - 5y)$

Page 154: **35.** $(7x - 11y)(11x - 7y)$ **37.** $(2a + 3b)(3a - 2b)$ **39.** $(5x^2 - 4y)(3x^2 - 2y)$ **41.** $(19x^3 - 2y)(x^3 - 3y)$ **43.** $(6x^2 - 5y^2)(2x^2 - 5y^2)$ **45.** $(6x^3 - 5y^3)(x^3 + 2y^3)$ **47.** $(9h^2 - 2b^2)(2h^2 + 3b^2)$ **49.** $(5a^3 + b^3)(7a^3 - 8b^3)$ **51.** $(2x + 7)(x + 3)$

Page 155: **1.** $x^2 + 10x + 25$ **3.** $9a^2 - 12a + 4$ **5.** $m^2 + 4mn + 4n^2$ **7.** $4a^2 + 12a + 9$ **9.** $m^2 - 18mn + 81n^2$ **11.** $(x + 7)^2$ **13.** $(y - 10)^2$ **15.** $(2x - 7)^2$ **17.** $(3a - 11b)^2$ **19.** $(0.2x + 0.3y)^2$ **21.** $11x + 1$

Page 156: **1.** $x^2 - 25$ **3.** $196a^2 - 1$ **5.** $x^2 - 49y^2$ **7.** $9a^2 - 25b^2$ **9.** $\frac{16}{49}c^2 - \frac{4}{9}d^2$ **11.** $(x - 4)(x + 4)$ **13.** $(3 + 5a)(3 - 5a)$ **15.** $(\frac{1}{2}x + \frac{1}{3}y)(\frac{1}{2}x - \frac{1}{3}y)$ **17.** $(13x + 5y)(13x - 5y)$ **19.** $(ab + c)(ab - c)$ **21.** $(1.1c + 0.3d)$

$(1.1c - 0.3d)$ **23.** $(13x + 2y)(13x - 2y)$

Page 157: **1.** $7(x + 5y)$ **3.** $5b(3a + 7c)$ **5.** $5(x + 2y)(x - 2y)$ **7.** $m(m + n^2)(m - n^2)$ **9.** $4y(x - 1)^2$ **11.** $3x(x - 2)(x + 1)$ **13.** $5a(xy + z)(xy - z)$ **15.** $5(m - 3n)^2$ **17.** $a(2a + 1)(a - 1)$ **19.** $3b(3 + 2b)(2 + b)$ **21.** $4xy(2x - y)(3x - 2y)$

Page 158: **23.** $3(y^2 + 2)$ **25.** $m(x - 2)(x + 1)$ **27.** $mn(n - m)$ **29.** $x(x + 1)$ **31.** $\pi r(2 - r)$ **33.** $3x^2y^2(9 - y)^2$ **35.** $(6xy + 7z)^2$ **37.** Prime **39.** $4(m^4 + 36n^6)$ **41.** $9(2a^2 + 5b^2)(2a^2 - 5b^2)$ **43.** $(11xy + 13z)(4xy - 5z)$ **45.** $(6ab + 7c)(7ab - 4c)$ **47.** Prime **49.** $6a^3(rs + 3t^2)(rs - 3t^2)$ **51.** $6y(n^6 + 36)$

Page 159: **53.** $(y - 4)(y + 1)$ **55.** $(y + 4)(y - 1)$ **57.** $(x - 7)(x - 2)$ **59.** $(x + 7)(x - 2)$ **61.** $(x - 7)(x + 2)$ **63.** $(y + 50)(y - 2)$ **65.** $(x - 10)(x - 2)$ **67.** $(y + 25)(y + 4)$ **69.** $(2 + y)(1 - y)$ **71.** $(x^2 + 1)(x + 3)(x - 3)$ **73.** $a(3 - a)$ **75.** Prime **77.** $9x^2(x^3 + 3x - 5)$ **79.** $y(x^2y + 1)$ **81.** $x(3x^3 + 11x + 13)$ **83.** $(4n - r)(2n + r)$ **85.** $[(e + f) + 3]^2$ **87.** $9(r + m)(r - m)$

Page 160: **89.** $(x + 2)^2(3xy)$ **91.** $x(x + 2)(x + 1)$

Page 161: **1.** a^4 **3.** $3x^2$ **5.** $-4x$ **7.** $15m^3n$ **9.** $21c^3d^2$ **11.** $ab + 1$ **13.** $-8xy^2 + 11$ **15.** $-14a^2 + 9a$ **17.** $5x - 6x^3$ **19.** $-11r^2 + 5r^4$

Page 162: **21.** $2 - b$ **23.** $7 + 4m$ **25.** $m^2 + 7m + 5$ **27.** $r^2 - 7r + 4 + \frac{24}{r-3}$ **29.** $3x^2 - 2xy + 5y^2$ **31.** $2a^2 + 3ab - 4b^2$ **33.** $(x - 2)$ hours

Page 163: **1.** $\frac{b}{4c}$ **3.** $\frac{5+x}{-2}$ **5.** $\frac{b+3}{-5}$ **7.** $\frac{4x-y}{x}$ **9.** $\frac{x-6}{x+1}$ **11.** $\frac{d}{d+2}$ **13.** 2 **15.** $\frac{a+2}{12}$ **17.** $\frac{5a+5}{a^3}$

Page 164: **19.** $\frac{2}{x-2}$ **21.** 1 **23.** $\frac{8y}{5x^2}$ **25.** $\frac{xy}{30z}$ **27.** $2a - 2$ **29.** $\frac{a^2-5a}{3a-9}$ **31.** $\frac{-1}{4}$ **33.** $\frac{-1}{2}$ **35.** 7

Page 165: **1.** $\frac{7}{2x}$ **3.** 11 **5.** $\frac{1}{y}$ **7.** $\frac{x+2}{x-2}$ **9.** $x + y$ **11.** $\frac{2}{x+y}$ **13.** $\frac{a^2-2b^2+3y^2}{a^2+y^2}$

Page 166: **15.** $\frac{4}{a-3}$ **17.** $\frac{2x+9}{4x+3}$ **19.** 45 mph **21.** $\frac{3a-2}{6}$

Page 167: **1.** $4x$ **3.** $14x^2y^2$ **5.** $4x^2 - 28x + 49$ **7.** $(x - 3y)$ or $(3y - x)$ **9.** $x^2 - 5x$ **11.** $4x + 16$ **13.** $2ab + b^2$ **15.** $3x - 6$

Page 168: **1.** $\frac{3x}{8y}$ **3.** $\frac{4-7x}{x^2}$ **5.** $\frac{7c-9b}{abc}$ **7.** $\frac{4a+9}{4a}$ **9.** $\frac{9d^2-27d+4}{6d^2}$ **11.** $\frac{x^2+1}{x}$ **13.** $\frac{x^2-3x-8}{x+2}$ **15.** $\frac{-3}{3x-5y}$ **17.** $\frac{9}{4a-3}$

Page 169: **19.** $\frac{3y-75}{y^2-25}$ **21.** $\frac{7y+5x}{x^2y-xy^2}$ **23.** $\frac{-3x-29}{x^3-2x^2-9x+18}$ **25.** $\frac{4x-6}{3x^3-4x^2-5x+2}$ **27.** $\frac{3x^2-19}{x^3-19x+30}$

Page 170: **1.** 5 **3.** $\frac{-6}{13}$ **5.** $\frac{5}{4}$ **7.** $\frac{-3}{4}$ **9.** 2

Page 171: **11.** $\frac{-2}{3}$ **13.** -7 **15.** undefined **17.** positive **19.** negative **21.** negative **23.** $m(\overline{AB}) = \frac{-5}{3}$; $m(\overline{BC}) = \frac{1}{3}$; $m(\overline{AC}) = \frac{7}{3}$ **25.** b and c

Page 172: **1.** $m = \frac{2}{5}$, $b = \frac{-3}{5}$ **3.** $m = 4$; $b = 7$ **5.** $m = -2$; $b = 0$ **7.** $m = \frac{-4}{9}$; $b = \frac{2}{3}$ **9.** $m = \frac{-1}{8}$; $b = \frac{5}{2}$ **11.** $3y - x = 9$ **13.** $y + 4x = 5$ **15.** $y + 2x = 0$ **17.** parallel **19.** perpendicular

Page 173: **1.** $3x - 4y + 15 = 0$ **3.** $4x - 3y - 11 = 0$ **5.** $20x + 9y + 1 = 0$ **7.** $x - y = 0$ **9.** $3x - y - 11 = 0$ **11.** $x + 2y + 5 = 0$ **13.** $3x - 5y + 15 = 0$ **15.** $5x - y + 7 = 0$ **17.** $3x + 2y + 8 = 0$

Page 174: **17.** $x + y = 0$ **19.** $2x + y + 4 = 0$ **21.** $x + 2y - 4 = 0$ **23.** $y - 2 = 0$ **25.** $3x - 4y - 13 = 0$

Page 175: **1.** $\frac{-9}{4}$ **3.** -2 **5.** $x + 5y - 23 = 0$ **7.** $m = 1$; $b = \frac{4}{5}$ **9.** $2x + 5y - 19 = 0$ **11.** $13x - 9y + 8 = 0$ **13.** $\frac{-1}{3}$; $\frac{12}{5}$

Page 176: **1.** $4\sqrt{2}$ **3.** $8x\sqrt{2x}$ **5.** $12ab\sqrt{2a}$ **7.** $15x\sqrt{2}$ **9.** $16x\sqrt{2y}$ **11.** $\frac{\sqrt{2}}{2}$ **13.** $\frac{\sqrt{21xy}}{3y}$ **15.** $\frac{2a^3\sqrt{30ac}}{5c}$ **17.** $\frac{\sqrt{6a}}{22}$ **19.** $\frac{xy^2\sqrt{z}}{z^2}$ **21.** $\frac{4\sqrt{5a}}{5a}$ **23.** 8

Page 177: **25.** $\frac{5\sqrt{a+b}}{a+b}$ **27.** $\frac{3\sqrt{x-y}}{x-y}$ **29.** $x\sqrt{3x}$ **31.** $\frac{x\sqrt{xy}}{11y^2}$ **33.** $\frac{x\sqrt{5xy}}{5}$ **35.** $\frac{8a^2\sqrt{6b}}{3b}$ **37.** rational **39.** rational **41.** rational **43.** irrational

Page 178: **1.** $11\sqrt{3}$ **3.** $2\sqrt{x}$ **5.** $4\sqrt{2}$ **7.** $5\sqrt{3}$ **9.** $2\sqrt{2}$ **11.** $-8x^2\sqrt{2}$ **13.** $7\sqrt{xy}$

15. $5b\sqrt{2ac}$ **17.** $\frac{2\sqrt{11}}{35}$ **19.** $\frac{3\sqrt{2}}{x^2}$

Page 179: **21.** $-7x\sqrt{2x}$ **23.** $30\sqrt{3}$ inches **25.** 15 meters **27.** 7 cm

Page 180: **1.** $30x$ **3.** $60x^2\sqrt{3}$ **5.** $-14a^4$ **7.** $25x\sqrt{2}$ **9.** $2\sqrt{6}$ **11.** $9x$ **13.** $2x^2yz^3$ **15.** 216 **17.** $140x^2\sqrt{3}$ **19.** $36a^2b^2\sqrt{c}$ **21.** $3\sqrt{2} - 3\sqrt{5}$

Page 181: **23.** 7 **25.** $\sqrt{3}$ **27.** $\frac{\sqrt{6x}}{2}$ **29.** $\frac{\sqrt{10y}}{2}$ **31.** $6\sqrt{2x}$ **33.** $\frac{\sqrt{15}+\sqrt{10}}{5}$ **35.** $54\sqrt{3xy}$ **37.** $11 - 5\sqrt{3}$ **39.** -67

Page 182: **1.** $\{-2, -7\}$ **3.** $\{6, 3\}$ **5.** $\{-7, 4\}$ **7.** $\{11, -5\}$ **9.** $\{0, -9\}$ **11.** $\{0, 15\}$ **13.** $\{8, 7\}$ **15.** $\{-12, -4\}$ **17.** $\left\{\frac{1}{2}, -\frac{1}{3}\right\}$ **19.** $\left\{\frac{3}{4}, -\frac{1}{8}\right\}$ **21.** $\{6, -9\}$ **23.** $\{-12, 5\}$

Page 183: **25.** $x^2 = 9x + 22$; $x = 11$ **27.** $w(w + 5) = 84$; width = 7 m; length = 12 m **29.** $n^2 - 27 = 6n$; $n = 9$ or -3

Page 184: **1.** $\frac{-1 \pm \sqrt{5}}{2}$ **3.** $-2 \pm \sqrt{10}$ **5.** $\frac{-3 \pm \sqrt{17}}{2}$ **7.** $\frac{6 \pm \sqrt{33}}{3}$ or $2 \pm \frac{\sqrt{33}}{3}$ **9.** $\frac{2 \pm \sqrt{5}}{2}$ or $1 \pm \frac{\sqrt{5}}{2}$ **11.** $\frac{3}{2}$ **13.** $h = 3\sqrt{2}$ cm; $b = 6\sqrt{2}$ cm

Page 185: **1.** $1\pm\sqrt{11}$ **3.** $-3\pm\sqrt{7}$ **5.** $\frac{-1\pm\sqrt{2}}{2}$ **7.** $\frac{-2\pm\sqrt{3}}{3}$ **9.** $\frac{3\pm\sqrt{13}}{4}$ **11.** 7 and 17 **13.** 8 or -9

Page 186: **1.** $-3, 7$ **3.** $\frac{3}{5}, \frac{1}{2}$ **5.** $3\pm\sqrt{14}$ **7.** $1, -\frac{2}{3}$ **9.** 6,0 **11.** 7,-4 **13.** $-\frac{3}{2}, -1$ **15.** $\frac{1\pm\sqrt{5}}{4}$ **17.** ±15

Page 187: **19.** $-5, 2$ **21.** $-\frac{1}{2}$ **23.** $\frac{2}{5}, -\frac{1}{6}$ **25.** $2 \pm 3\sqrt{2}$ **27.** $-\frac{3}{2}$ or 3 **29.** $\frac{1}{3}$ or 5 **31.** 8 in. and 15 in.

Page 188: **1.** c **3.** b **5.** b **7.** a **9.** b **11.** b **13.** a **15.** $x^2 + 15^2 = 25^2$; $x = 20$ ft

Page 189: **1.** 5 mi **3.** 951.39 mL or 950.5251 mL **5.** $w = 7$ ft; $\ell = 29$ ft

Page 190: **1.** c **3.** d **5.** d **7.** b **9.** d **11.** d **13.** a

Page 191: **15.** d **17.** c **19.** c **21.** b **23.** a

Page 192: **25.** d **27.** c **29.** d **31.** b **33.** a **35.** d

Page 193: **37.** a **39.** b **41.** $\frac{6y + 3x + 4xy}{x^2y^2}$ **43.** $\frac{2\sqrt{3x^3 + 9x}}{x^2 + 3}$ **45.** $\frac{-a^2 - 20a + 10}{(a + 3)(a - 5)(a - 2)}$ or $\frac{-a^2 - 20a + 10}{a^3 - 4a^2 - 11a + 30}$ **47.** 14 dimes; 6 nickels **49.** 16 cm, 30 cm, 34 cm